PENGUIN BOOKS

# BHOPAL : THE LESSONS OF A TRAGEDY

Sanjoy Hazarika was born in 1954 in Shillong, Northeast India. He schooled at St. Edmunds, Shillong, and later studied journalism and printing in London.

Mr Hazarika has worked with several magazines, newspapers and news agencies in India and, since 1981, has been the resident *New York Times* correspondent in New Delhi. He was selected for the *New York Times* Publisher's Award in December 1984 for his reporting on Bhopal. He has reported events in Sri Lanka, Afghanistan and the United States and has also travelled extensively throughout Southeast Asia, Europe, North America and the Indian subcontinent.

Sanjoy Hazarika lives in New Delhi with his wife and daughter.

# BHOPAL

## THE LESSONS OF A TRAGEDY

Sanjoy Hazarika

PENGUIN BOOKS

Penguin Books (India) Private Ltd, 72-B, Himalaya House
23 Kasturba Gandhi Marg, New Delhi-110 001. India
Penguin Books Ltd, Harmondsworth, Middlesex, England
Viking Penguin Inc, 40 West 23rd Street, New York,
New York 10010, U.S.A.
Penguin Books Australia Ltd, Ringwood, Victoria, Australia
Penguin Books Canada Ltd, 2801 John Street, Markham,
Ontario, Canada L3R 1B4
Penguin Books (N.Z.) Ltd, 182-190 Wairau Road,
Auckland 10, New Zealand

First Published by Penguin Books India 1987


Made and printed in India by
Ananda Offset Private Limited, Calcutta
Typeset in Times Roman

*For Minu and Meghna*

# Contents

# A Note On Style

I have used abbreviations and various names for several of the institutions mentioned in the book.
Thus, Union Carbide Corporation is referred to as the corporation, Union Carbide or simply as Carbide. Its Indian subsidiary, Union Carbide India Limited, is however described as UCIL in second references.
The Indian Government is called New Delhi, the Central Government or just India. India refers thus to both the country and its government, depending on the context.
Multinational Corporations are MNCs, transnational corporations or multinationals.

S.H.

# Acknowledgements

Some of the ground work for this book was done during the three weeks of reporting I did for the *New York Times* in December 1984 at the time of the disaster. But in the past three years, I have personally conducted more than 250 interviews with those in any way connected with the disaster in New Delhi, Bhopal and Bombay. Much of the research and findings are my own. However, I have used published material for some sections of the book and I wish to thank the authors, publishers and others for permitting me the use of material from their books: Christopher Tugendhat, author of *The Multinationals* (published by Eyre and Spottiswoode Publishers Ltd, Methuen, London); John Wiley & Sons Inc. of New York, publishers of *Multinationals under Fire: Lessons in the Management of Conflict* (authors: Thomas N. Gladwin and Ingo Walter); the estate of Rachel Carson and Hamish Hamilton, publishers of *Silent Spring;* excerpts from *Silent Spring* by Rachel Carson have also been reprinted by permission of Houghton Mifflin Company; to James Erlichman, author of *Gluttons for Punishment* (1986), for permission to use material from pages 28, 29 and 126, which have been reproduced by permission of Penguin Books Ltd.

I also wish to acknowledge permission granted by the International Organization of Consumers Unions, Penang, Malaysia, publishers of David Weir's *The Bhopal Syndrome* to use material from the book; the European Community office in New Delhi for providing details of the Seveso Directive and debates about the incident; The Institute for Food and Development Policy at San Francisco, publishers of *Circle of Poison* by David Weir and Mark Shapiro.

There are those, too, I must thank but who would rather not I named them, especially officials in the Government of India and the Governments of Madhya Pradesh and Maharashtra who gave me detailed briefings, made documents available and provided other kinds of assistance.

Similarly, some familiar with the other point of view, namely Union Carbide's, have been helpful in explaining the corporation's views, especially on legal matters. I am particularly grateful to Vijay Gokhale, the chairman and managing director of Union Carbide India Limited, who talked with me at length in Bombay. Others I have met include Jagannath Mukund, then UCIL's works manager at Bhopal, lawyers for the company and other employees.

The UCIL trade union, and R.K. Yadav, in particular, helped with information about the incident and the events preceding and following it. Other workers who provided key details include Suman Dey, T.R. Chauhan, A.K. Khamparia and Shakeel Qureshi, the then shift supervisor in the unit that leaked.

To those who discussed the issues studied in this book, who read parts of the manuscript and made valuable suggestions and corrections, I express my gratitude. They include Vinayak Shah, Russi M. Lala of the Dorabjee Tata Trust, Arun Kumar of Jawaharlal Nehru University, Salil Tripathi, Raj Kumar Keswani of Bhopal, Dr. Heeresh Chandra and Dr. N.P. Mishra of Bhopal's Hamidia Hospital, Dr. Vulluri Ramalingaswami, then director-general of the Indian Council of Medical Research, Dr S.R. Kamat and his team from Bombay's King Edward Medical College, Dr. S.G. Goyal of the Indian Institute of Public Administration, the Voluntary Health Association of India, especially Dr. Meera Shiva, Jitendra Tuli of the World Health Organization office in New Delhi who put together documents and reports for my use; the Environment Services Group and Thomas Mathew for lending me documents (and their grace in allowing me to keep the stuff for long months), Ravi Chopra, friend and environmental activist, and Kalpana Sharma of the *Indian Express*.

Particular thanks are due to Rakesh Gandhi, managing director of Raba Contel, distributors for Apple Computer systems in South Asia, for the use of computer facilities at their office. Kalpana

Dubey, Raba Contel's development manager, was a constant, courteous help, despite the pressure of other demands.

To the Centre for Science and Environment, New Delhi, I wish to express my thanks for the use for their excellent library. I am particularly grateful to Anuradha for her assistance in locating material. I also wish to acknowledge the use of material from the Centre's *State of the Environment Report* for 1984-85. It was very helpful in preparing the chapter on India, particularly the study of occupational health hazards. The details of the Minamata disaster are from a report by Dr. Masazumi which was made available by the Centre.

Rajendra and Rupal Desai generously let me use their computer facilities in Bombay. Ramu Pandit, secretary-general of the Indian Merchants Chamber at Bombay, was a great help in providing details of Indian industry's response to the Bhopal tragedy. I enjoyed my long conversations with Dr. Nilay Choudhury, former chairman of the Central Board for Control of Water and Air Pollution. I have been constantly encouraged in reporting on Bhopal by the *New York Times,* especially by Warren Hoge, former foreign editor, and Annie Zusy, who maintained an abiding interest in the issue. I know that many sources spoke to me and helped me largely because of my association with the *Times.*

To Mike Kaufman of the *New York Times,* who picked me for the New Delhi bureau, I owe a special debt of gratitude. Bill Stevens (who was bureau chief during 1984) and I shared some of the most exciting and tragic moments of that year. He remains a real friend and extremely supportive. Steven R. Weisman, the present bureau chief, has also been supportive.

David Davidar, my editor, showed great patience, sensitivity and skill.

But if there is an award for patience, it should go to Minal, my wife, who's had to handle a preoccupied husband and an energetic four-year-old daughter virtually on her own, all these months. To them I dedicate this book, with the knowledge that life with a reporter, especially one working for a Western news organization, can never be soothing.

I know that without her and Meghna, this book would never have been written. Why Meghna? For her joyful and vibrant nature is a compelling contrast and reminder of the opposite: the hurt, the fear, the panic and the agony of Bhopal's children — dying in their homes, streets and hospitals. It is a painful memory that endures.

Sanjoy Hazarika
*New Delhi*
*November 1987*

# Introduction

*Without chemicals, life itself would be impossible.*

That glib catch-phrase was coined by the Monsanto Company, a major chemical producer which features in the *Fortune 500* list of the world's biggest industrial corporations.

Yet, a top Monsanto official was sufficiently chastened by the Bhopal disaster to remark afterwards that the American public 'simply does not have faith that the chemical industry will regulate itself in the public interest.'[1]

Multinational corporations, or MNCs as they are popularly known, like Monsanto and Union Carbide, from whose Indian subsidiary a fatal gas leaked in December 1984 at Bhopal, killing and injuring thousands, dominate the world chemical trade. They also control other important sectors of industrial manufacture and agricultural production. Charles Pearson of Washington's World Resources Institute has estimated that the top eight MNCs control sixty per cent of the non-Communist world's production of pesticides.[2]

The UN Centre on Multinational Corporations in a recent report estimated that there were about 11,000 MNCs with more than 82,000 foreign subsidiaries and affiliates; 21,000 of these foreign units were located in the most backward countries; thirty-six per cent had parent firms in the US, twenty-seven per cent were from Britain, seven per cent from France and six per cent from West Germany and Japan. The document says that parts of the Western Hemisphere, such as Latin America, have a concentration of forty-seven per cent of all multinational affiliates in developing countries. Another twenty-eight per cent are located in South and East Asia, twenty-one per cent in Africa and the rest in West Asia.

Multinationals, therefore, are an inescapable part of our life today. They cannot be wished away, even after a disaster like Bhopal. They form an important part of the world economy and conduct business in its remotest corners, with a combination of pressure, persuasion and packaging. They are incredibly powerful and some of their budgets exceed the Gross National Products of many nations.

Exxon, the petroleum giant which was listed by *Fortune* magazine as the top US corporation in 1984,[3] had annual sales of $88.5 billion that year. Its assets were estimated at $62.9 billion[4] and its sales[5] were worth more than three times the GNP of Libya ($25.1 billion),[6] four times that of Chile ($21.8 billion)[7] and nearly forty times that of either Nepal or Nicaragua[8] (each was assessed by the World Bank as having a GNP of around $2.6 billion).

In fact, Exxon's sales and assets together easily surpassed the individual GNPs of Switzerland, Saudi Arabia, the Netherlands and fell just short of that of Australia.[9]

Union Carbide, then a lowly thirty-seven on the *Fortune 500* list (it dropped to thirty-nine in the 1985 listings after the gas leak) had sales of $9 billion and assets of $10.3 billion in 1983.[10]Its sales alone were good enough to beat the GNPs of Burma ($6.5 billion),[11] Kenya ($6.4 billion)[12]and El Salvador ($3.6 billion).[13]

This financial power has sometimes led to MNCs being a law unto themselves. Several instances have been recorded of efforts by MNCs to evade or change laws and defy international conventions.

'Screw them...unless we are really wanted, to hell with them.'[14] With those caustic remarks, Zoltan Merszei, a former Chairman of the Dow company, summed up his personal attitude and perhaps that of his tribe towards governments that place hurdles in the path of corporate growth. Dow has faced litigation in many parts of the world on charges of harming the environment.

MNCs are often seen at their best or worst (depending on who's doing the looking) when fighting to boost or save company reputations, investments and profits. Thus, International Telephones and Telecommunications' chief executive, Harold S. Geenen, boasted in 1962 that the company had 'met and surmounted every device employed by governments to encourage their own industries and hamper those of foreigners, including taxes, tariffs, quotas, currency restrictions, subsidies, barter arrangements, guarantees, moratoriums, devaluations...yes, and, nationalizations.'[15]

Some of the biggest companies have often broken the law to suit their own purposes. For example, Lockheed, the American aircraft manufacturer, made the headlines in the 1970s when it paid millions of yen to Kakeui Tanaka, the then Japanese Prime Minister, and other officials to favour Lockheed in trade deals. Tanaka resigned when details of the operation came to light.

Sometimes, multinationals move beyond bribes to actually assisting their governments overthrow 'hostile' regimes. The destabilization of governments in Iran and Chile are the two best documented examples — although nearly 20 years separate the two events.

In Iran, the United States acted against nationalist Premier Mohammad Mossadegh in 1953 after he nationalized the Anglo-Iranian oil company. Mossadegh was overthrown with the help of the US Central Intelligence Agency.[16]

ITT and the Kennecott and Anaconda corporations are said to have worked with the United States and the CIA to topple a Marxist regime in Chile in 1973. 'Covert American activity,' said a US Senate report, 'was a factor in almost every major election in Chile between 1963 and 1973.' In several instances, the United States' intervention was significant.[17] The *New York Times* reported in 1976 that a US Federal grand jury had ruled that the CIA and the ITT colluded to fabricate and coordinate statements before a Senate investigation of ITT's role in the affair.[18]

By 1970, President Salvador Allende had begun slashing the control of the MNCs, and that of the US Administration, which extended to vast sectors of the Chilean economy. In their book, *The United States and Chile; Imperialism and Overthrow of the Allende Government,*[19] James Petras and Morris Morley say that by the end of 1970, the country's chief foreign exchange earner, copper, was virtually run by mining giants Anaconda and Kennecott. The manufacture and distribution of pharmaceuticals, office equipment, advertising, automobiles, radio, television and copper fabrications, were almost totally in the hands of the MNCs.[20] Unsurprisingly ITT, Anaconda and Kennecott battled

Chilean plans to nationalize major industries. Then the CIA got into the game which culminated with the coup that overthrew Allende.

Thus, MNCs can and do intervene decisively in national and international affairs, influencing policy, politicians and worldwide prices.

Another dimension to multinationals that invites censure is the fact that some of the biggest multinationals are leaders in arms manufacture and so have an assumed interest in the militarization of mankind.

Yet, MNCs are not always in conflict with local officials nor can they be dubbed warmongers. That would be the simplistic view, for they are also involved in vital medical, chemical and technological research critical for man's survival.

The developing world in particular cannot do without the skills and resources of the MNCs. Nor can MNCs do without the markets of the developing world, which are becoming accepted as necessary to global development. The issue is whether a more balanced relationship can be worked out, one which takes care of the concerns of either side.

But the Bhopal disaster was adequate notice that in the key area of safety and environmental protection, MNCs and their subsidiaries were not doing enough to protect their own workers or the communities around them. It also pointed to the inherent dangers faced by Third World Governments which allow high-technology-oriented multinationals to do business in their countries. Indeed, Bhopal underlined the need for MNCs, local industry and government to be knowledgeable about each other's function. There was little point, the incident showed, in the various parties regarding one another as adversaries. Cooperation, the sharing of information, the devising of new anti-pollution laws and the implementation of existing ones would have to become a mutual imperative.

Twenty-five years ago, Rachel Carson wrote in *Silent Spring*[21] of a world 'where no birds sing,' the possible result of indiscriminate pesticide use. She was making the point that ours is a small planet and that we must preserve it or perish.

Carson's message appears in danger of being forgotten today in the chemical industry's scramble for profits. Experts say that most pesticides are inadequately tested for possible human carcinogens or cancer-producing compounds.

Officials at the Environment Protection Agency in the US, one of the better staffed and funded government agencies of its kind anywhere, say the retesting of existing pesticides in the United States alone, and the researching of their possible effects on human health, would take another twenty years. According to the Chemical Abstract Service, a register of chemicals produced worldwide, about 7.3 million chemical substances have been registered in the past 20 years.[22] More than 1.5 million substances were in use before the register began publishing.[23] From these approximately nine million chemicals, some 70,000 chemical products are on the market.[24] About 1,000 new chemicals enter commercial use every year, said another report in January 1985.[25] Many of these chemicals are exported.

According to a World Health Organization report, India produces about 70 per cent of the total insecticide production of Asia and Africa.[26] Also, the developing world as a whole accounts for about twenty per cent of world pesticide use.[27] Brazil, India, Mexico, South Korea, Malaysia, the Philippines and Thailand are the major users.[28] Pesticide related deaths were placed at about 10,000 every year in developing countries.[29] Another report estimated that 1.5 to 2 million suffer from pesticide poisoning,[30] caused by the indiscriminate spraying of crops or vegetation, failure to wear adequate protective clothing and widespread general ignorance of the dangers of the chemicals being used.

Officials in developing countries defend the extensive use of pesticides saying that they are critical to agricultural self-reliance and the reduction of food imports and subsidies. Indira Gandhi, the late Indian Prime Minister, is quoted approvingly by these officials as once saying that 'poverty is the greatest pollutant in the case of developing countries.' This wilful ignoring of the dangers of pesticide use is alarming.

The high use of DDT and BHC in developing countries has led to dangerous levels of these products in the bodies of the inhabitants of the poor world. (For the most part, these substances are banned in industrialized countries). Human breast milk in China and India shows several times the median levels of DDT and BHC in developed nations;[31] both pesticides can cause an array of health problems. Besides, the use of such chemicals pollutes the entire food chain and the environment.

An unfortunate aspect of the whole problem is that when some Western MNCs were forced by tough laws and environmental pressure to move their hazardous operations in developed countries, they relocated in countries with poor pollution control laws. Martin Abraham, an editor with a consumer research organization in Malaysia, notes that Japanese MNCs are still relocating their pollution-intensive industries, via foreign investment, in developing countries.[32] 'In fact, it is estimated that between two-thirds and four-fifths of all of Japan's pollution-intensive foreign investments have been located...mainly in Asian and Latin American countries.'[33]

What big industry needs to realize is that this sort of shortsighted flouting of environmental imperatives cannot go on forever. The resources of the world are being depleted at an alarming rate, creating famines and floods. This is clearly one area where industry,(especially MNCs),with its resources and managerial skills, can cooperate with host governments. It makes good sense and good business these days to invest in protecting the environment, for industrial pollution is not just a problem of the developing world .

The potential hazard from industrial pollutants is staggering. A US Congressional committee, studying pollution by American chemical companies, reported in March 1985 that sixty-seven companies had listed 204 hazardous chemicals which they were using.[34] Henry A. Waxman, the Democratic Representative heading the investigation, declared that no US Government agency had ever attempted to compile a national inventory of toxic chemical emissions[35] (methyl isocyanate, the deadly gas that escaped from the Union Carbide plant at Bhopal, was not even listed as a hazardous air pollutant at the time). The study found that many chemical plants in that country routinely released tons of untreated hazardous substances into the air. An official at an Exxon plant in Louisiana told the committee that the plant released more than 560,000 pounds of benzene every year.[36] Benzene is known to cause leukemia in human beings.

Later that year, the EPA released a list of 402 toxic chemicals stored or otherwise used in American plants which could endanger communities in the event of spills, accidents or leaks.[37] It was also reported then that more than 550 companies in almost every American state handled these substances almost routinely.[38] Acid rain, another direct result of industrial pollution, has been killing fish and other marine life in Europe and North America for years. Indeed, the West, despite its technological edge, has been hit time and again by environmental disasters. The German Department of Agriculture says that the total percentage of forests damaged by acid rain rose in that country from eight per cent in 1982 to fifty per cent in 1984. The Minamata disaster of the 1950s and 1960s in Japan, the pre-eminent Asian nation in terms of industrialization, was one of the most tragic examples of the danger of untreated industrial effluents. Hundreds of Japanese were crippled by mercury poisoning after they ate river fish contaminated by effluents from the Chisso Company.

The accident in July 1976, at Seveso near the Italian city of Milan,[39] is possibly the best-known industrial tragedy in the West. The disaster had much in common with Bhopal. Both companies were subsidiaries owned by MNCs; both disasters involved the leakage of a toxic chemical; in both cases, the management delayed giving crucial information to authorities and the neighbouring community. Local officials in both cities were casual in their attitudes toward a hazardous industry and the implementation of safety laws.

In the Seveso incident, Givauden, a subsidiary of Hoffman-La Roche, the Swiss chemical MNC, spewed a massive amount of dioxin across several communities neighbouring the plant.[40] Most residents paid little attention to the pungent smell of the gas and the company even assured local officials that there was nothing to fear. The plant resumed operations.[41]

Within a few days of the leak, pets began to bleed at the nose and mouth and die. Vegetation withered and thousands of fowl and domestic livestock collapsed and died. Residents began to complain of blistering skin, diarrhoea, headaches, dizziness, kidney and liver pains. A week after the disaster, the company disclosed, for the first time, that dioxin, a substance used as a defoliant by US armed forces in Vietnam and also as a bactericide (used for bodycare, according to the producer), had been released. More than 800 people were evacuated and top company officials were arrested.[42]

The accident triggered off an outcry against MNCs and drew attention to the poor controls that even industrialized nations place on major corporations. Later Givauden admitted that it had set up the plant near Milan because strict Swiss environmental legislation had forced it to locate in a country with laxer laws. A check of chemical plants around Milan after the disaster showed up 300 firms which had been violating regulations.[43]

All this goes to show that risk control systems need to become more sophisticated and efficient. Part of big business' resistance to regulation stems from a fear of dropped profits and being edged out by less scrupulous competitors. One of the major problems about developing safeguards, industry claims, is its high cost. Capital would be diverted to unproductive, pollution-control investments; company funds would go into the design, operation and maintenance of pollution-control equipment. But there can be no short-cuts (or reduced rates) to safe work environments. MNCs heed to pay particular heed to this if they aren't to be seen in host countries as the baddies — ready to dump toxic materials on unsuspecting, poor societies at the first opportunity. Currently a majority of 'high-risk' MNCs are perceived as deliberately seeking pollution havens (as we've seen earlier this is not mere fanciful theorizing given the number of well-documented cases of MNCs indulging in such practices) in countries with casual attitudes and light laws governing industrial pollution. Some more examples to show the direct impact irresponsible industries have on the environment would be in order. David Weir, author of *The Circle of Poison,* cites the example of the American company Velscicol. The firm's plant at Raport, Texas, shut down in the 1970s after workers there reported major nervous disorders after handling phosvel, an insecticide.[44] The chemical was banned then but Velscicol continued to export it to Colombia, and other countries.[45] In 1971, exposure to phosvel killed more than 1,000 water buffalos and an unknown number of peasants in Pakistan.[46]

Other products, banned or restricted in the United States and other Western nations, like aldrin, dieldrin, heptachlor, BHE, methyl parathion and melathion are sold to developing countries such as India, Costa Rica and the Philippines. All this is documented fact and cannot but reflect badly on MNCs, especially the chemical and drug companies.

However, it is true that the advanced technology and superior assets of MNCs often help them build better and safer plants then local industry. One study on transnational enterprices reported that 'there is little evidence that (they) behave worse than local industries, and that, in fact, there is a good deal of direct and circumstantial evidence that MNCs have a better record.'[47] There is much truth in this. In the developing world there is no doubt that the worst offenders are government factories, local private industry and illegal manufacturing units located in the courtyard, basement or room of a home or in a crowded neighbourhood.

Encouragingly, MNCs appear to be moving toward a measure of self-regulation in the manufacture, storage and distribution of hazardous substances. This process has been accelerated by events such as the 1972 UN Conference on the Environment at Stockholm. More than 100 countries set up statutory bodies to assess the ravages of industrialization on their natural resources and began anti-pollution legislation. In the United States there are also groups such as the Pharmaceutical Manufacturers Association, the Chemical Manufacturers Association and the Centre for Chemical Plant Safety which press for safety measures. The last group was set up in response to the Bhopal crisis and has issued hazard-evaluation procedures for the US chemical industry. Another body, the European Council of Chemical Manufacturers Federation, has urged technology transfer to ensure that the degree of safety is the same worldwide. Gradually the laws laid down by regulatory bodies are becoming comprehensive. Unfortunately, there is little documentation to show whether the various codes and laws are heeded and implemented. As has been mentioned earlier, it is in the common interest that the various controls and safeguards are adopted. Responsible industrialization is called for if MNCs are to benefit the countries in which they operate and themselves. Rajiv Gandhi, the Indian Prime Minister, was stating an obvious truth when he said in 1985: 'The risks facing the developing countries in the area of higher technology has increased manifold but international procedures of surveillance of policies and practices of transnational corporations have yet to acquire form and content.'[48] These risks have to be controlled immediately if more Bhopals are to be avoided.

Like all businesses everywhere, multinational corporations are primarily concerned with profits. However, it is inalienable that the world needs their expertise and skills, their resources and leadership, for with their emphasis on excellence, state-of-the-art facilities, high salaries and strong economic profiles (that attract some of the best minds around) they are extremely strong catalysts of global development. Indeed, Christopher Tugendhat describes, in his book *The Multinationals,* the phenomenal growth of multinationals as one of the most dramatic developments of this century. He argues that governments, especially in the developing world are coming under increasing pressure to formulate policies regarding MNCs for they often represent the unofficial, national interests of major industrialized nations like the United States, Britain and Japan.[49] Multinationals have aspects about them that are unique. Given their strength, size, financial and political clout, multinationals not only make their own rules but inspire loyalties among their employees, that are often extranational. This means that for senior MNC executives the country in which they work is almost incidental. Tugendhat underlines this aspect when he writes that the key element in the MNCs' success is their business executives' commitment to headquarters, the parent firm, over and above loyalty and service to the subsidiary. Analysing this relationship, he compares it to the chain of command in an army: 'Brigadiers and battalion commanders are powerful and important men. Much is left to their discretion and initiative. Their advice is sought by the general and taken into account when the plan of campaign is drawn up. If their spirit of judgement fails the army will be undone. But the limits of their authority are set at headquarters, and can be increased or diminished as the general decides. They take their decisions and formulate their plans within the context of his strategy, and in the knowledge that he can sack or promote them as he wishes. The same applies to a multinational company.'[50]

The objective is clearly to maintain executive and corporate control. There is nothing exceptionable about this. A business executive is eventually judged by the functioning of the units and managers under him and their ability to produce profits. This, incidentally, is an important aspect of the charges that Union

Carbide Corporation has defended itself against—that as a majority stockholder in the Indian subsidiary, it knew more than most what was going on and hence needed to shoulder more responsibility than it had shown itself willing to accept.

Local subsidiaries are given major powers to hire and fire, build, expand, promote, experiment, negotiate, expedite, invest and so on. An element of decentralization exists. But the subsidiary or affiliate cannot take the really big policy decisions on its own. The parent company is normally consulted at every step and local employees know their careers and promotions rest on the kind of impression they can make on their distant bosses.

The parent firm retains control by various means, most frequently by using a central command system with enormous funds, and several specialist planners at its disposal. Subsidiary managers do often try to suit their products to local conditions but, as Tugendhat says, the local company 'has an overriding extraterritorial commitment to its parent company.'[51]

This transnational loyalty is not new.

The growth of nationalism and, concomitantly, industry in the eighteenth and nineteenth centuries preceded the advent of today's MNCs. Nations, eager to industrialize, developed markets and opened the way for joint collaborations with industry from other nations. The instrument of tariffs was used to reduce imports and increase domestic production as local companies aligned with large international firms, sharing their technology.

Soon companies from the United States and Europe were exploiting the vast mineral and other resources of Asia, Africa and Latin America. By the end of the nineteenth century, many MNCs were familiar names in colonial outposts and elsewhere. The growth of MNC subsidiaries really boomed after World War II with the emergence of dozens of newly-independent nations that wanted to use western technology.

However, the seeds of transnational corporations were first sowed throughout the world with the advent of empire. Before they were brought under foreign rule, the nations of Africa, Asia and Latin America were basically self-sufficient with a simple trade mode of production.

The arrival of the colonialists meant the conversion of 'native' economies into export-oriented entities and a corresponding involvement in international trade. The usual equation was raw material from the colonies in return for finished goods made from that very material by the colonizers (indeed, the British East India Company was the first MNC the Indian subcontinent had ever seen). Naturally enough, this wrought havoc on the traditional economic and social traditions of the native states. Another outcome was that wealth and resources were concentrated in a few hands, increasing the great disparities that had always existed. This sort of imbalanced industrialization did not vanish when many colonial states became independent. Many Third World economists pushed the idea that a rapid increase in per capita income and GNP was the only way the newly independent countries could become self-sufficient. The way to achieve this was through an enlarged industrial capacity linked with greater agricultural production and public development works such as schools, roads and service industries. These would lead, it was hoped, to eventual self-sufficiency. But such development needed large, liberal doses of foreign investment and technology. The multinationals reappeared in greater numbers than before

Decades after independence, most nations which won freedom in the 1940s and 1950s are still impoverished. Backwardness and hunger persist. Self-reliance remains a distant dream for most countries, so deeply in debt are they to international lending institutions and other nations. A few, like India, have shown that stumbling, awkward steps at reducing dependence on foreign technology and capital can pay dividends—to an extent.

An interesting aspect of the developing world's 'aversion' to Western MNCs is that many poor countries have increased trade and cooperation among themselves. Big industrial groups from developing countries have been quick to exploit the opportunities and have emerged as MNCs in their own right. Louis T. Wells Jr. reported in a 1982 study that he had identified nearly 2000 subsidiaries of such transnationals headquartered in the Third World.[52] The majority of the subsidiaries, Wells noted, 'are located in countires that are clearly less developed than the countries in which the parent firm resides.'[53]

Third World Governments often prefer MNCs from a developing country because they operate more cheaply, are labour-intensive, select areas which are priorities for the host nation, and are less of a political and economic threat. But there are ancient suspicions, political and economic dislikes which also create hurdles to continuing, effective South-South cooperation. But it must be emphasized that developments such as the evolution of Third World MNCs are still marginal; Western MNCs still maintain extremely strong positions in areas such as the processing, distribution and marketing of agricultural products.

According to the World Resources Institute, MNCs control an estimated 80 per cent of agricultural commodity exports and have a virtual monopoly on major new technology which the developing world needs.[54] They thus maintain a role, if no longer major direct control, in the exploitation of natural resources and are partners and catalysts in technology transfer, processing, distribution and marketing.

In the final analysis, the role of the MNCs in advancing industrialization cannot be condemned. Yet it is also a good thing that people around the world are voicing fresh concern about the unknown frontiers that man is exploring, of forces he has tapped but not conquered, of knowledge he has gained but not mastered.

The world remains, as one writer put it, 'by turns happy and uneasy' with the technologies it cannot do without.[55] 'Though manifestly imperfect, the technologies have been accepted because their benefits seem to have outweighed the risks,' said John Noble Wilford, a science writer, in the *New York Times.*[56] Yet, death and disaster in any form lead to reflection on the price that the world pays for progress.

MNCs like Union Carbide are leaders in the whole business of development and progress. With the world's natural resources rapidly diminishing, MNCs, local industry, environmentalists and governments have a joint responsibility to save and sustain the earth. However it is a truism that neither governments nor MNCs nor even local industries can be expected to show generosity and nobility just because the world's ecosystems are threatened. That is why the broadening of laws to control the growth of hazardous industries and the dangers they represent is crucial. It is here that a 1986 verdict by the Supreme Court of India — applauded by activists and opposed by industry—fixing 'absolute and non-delegable liability,' for mishaps involving dangerous substances, on company managers, gives direction to the fight to regulate high risk industry. Such a ruling can be used to curb industrial callousness anywhere; it can also be argued that similar liability should be imposed on government agencies which are to check and regulate these processes. A commitment to safety must be enforced on all sides.

This book deals primarily with the Bhopal disaster —the world's worst industrial tragedy. The most obvious question the tragedy poses is whether there are more areas in India and other countries of the developing world which are potential disaster areas. The answer to that is 'Yes' for, to the poor world, economic development means more than almost anything else. But the developing world's indiscriminate welcome of any means to enhance its economic prosperity cannot be taken advantage of by careless multinational corporations or large indigenous industrial houses. Their responsibility for safeguarding the environment in which they operate and the people who work for them cannot be stressed enough. Governments in the developing world need to realize that economic prosperity cannot be put above everything else.

Yet, this is not an anti-technology or an anti-multinational book. I have only asserted a fundamental belief—that without restraint and compassion, technology and business can be devastatingly destructive to people and the environment in which they exist. We must learn to live with progress and use it to enrich our world, not destroy it.

Progress does not mean repetitions of tragedies like Bhopal. Callousness and greed know no frontiers, recognize no governments, respect no racial differences, observe no trade or national barriers. If governments and industry do not understand the lessons of Bhopal, then such disasters will continue to happen. Without a combined effort, our lakes and rivers will continue to be poisoned, our forests destroyed, our cities choked with the homeless and hopeless, our mineral wealth laid waste, millions of aboriginal people displaced and deprived and industrial and ecological disasters will become commonplace. The result will be a nightmarish world that we would have brought upon ourselves.

The choices are clear. It is a question of whether big industry, government and public interest groups have the will and the foresight to cooperate and prevent such a grim future. If they do not, our children will not inherit the earth. They will be left with dust bowls instead of rice fields, effluents instead of pure air and water—a legacy of pain and hunger in a world they may never conquer.

# Prologue

*Who or what caused the Bhopal disaster may not be conclusively proved for years. But there are essentially two separate scenarios that vary widely from one another. Each has its merits and lacunae as we shall see in the course of this book. The first scenario is structured out of the accounts of workers at the Union Carbide plant, activists, government scientists and journalists. It is dramatic and chilling and purports to show how negligence can lead to tragedy. The other scenario, drawn by Union Carbide's management and lawyers, is equally frightening for it depicts a saboteur determined to avenge an apparent slight. The second theory has had few supporters this far. But Union Carbide believes it has compelling evidence based on scientific tests, that will rule out every other cause for the disaster except sabotage by a specific employee.*

# SECTION I

# Out Of The Evil Night

About half-an-hour after midnight on 2 December 1984, Suman Dey,[1] a worker at the Union Carbide pesticide plant in Bhopal, stood on top of a set of shuddering concrete slabs. Below him, roofed by the concrete slabs, were three enormous stainless steel cylinders containing a chemical compound, methyl isocyanate., It was the chemical which was responsible for the vibrations for at that moment an intense reaction involving methyl isocyanate, water and other contaminants was under way in one of the tanks below his feet. The chemical reaction soon made the concrete hot to the touch and Dey, a tall, gentle-faced, thirty-two-year-old Bengali, who had worked at the plant for four years, was forced to leap to safety. He was just in time, for the next thing he knew the top of the concrete shield split with a loud crack. In a few moments, the lethal contents of Tank No. 610 would flow over the city of Bhopal.

Across the road from the insecticide factory, the residents of the slum colonies slept. They lived in wooden shacks held together with tarpaulin, rope and sacking, the roofs kept in place with stones. They had no proper toilets or roads and knew nothing of the danger from beyond the factory walls. That ignorance was shared by many living further south in the crowded old quarter of Bhopal. As with old quarters of most cities in India, it belonged to another, more deprived age, with narrow, twisting lanes, and tightly packed two-and three-storey-high brick and concrete buildings clustered on either side of the road.

Unlike the people in the slums, Dey knew that the methyl isocyanate (the plant workers called it MIC) which was turning from liquid to gas in the tank, was a killer.

Sick with fear and horror, Dey, who worked at the Union Carbide installation as a plant operator, watched a lethal, cream-coloured cloud leave the mouth of a stack connected to the leaking tank, surge higher into the air and then, propelled by a gentle night wind, sink toward the squalid shanty-town of Jayaprakash Nagar. The slum was separated from the plant by only an eight-foot-high wall and a bare fifteen feet of road. The gas, which was heavier than air, hugged the ground as it moved into the slum, nudged by the breeze It floated to the railway workers colony, the railway station, the old city's neighbourhoods of Kazi Camp and Chola Kenchi and the newer settlements of Sindhi Colony and Hamidia Road.

A stream of poison gas swamped the factory also, spreading a general panic among the workers and reducing vision. Workers grabbed oxygen masks. V.N. Singh, a worker in the MIC unit, and a member of the rescue unit, activated the factory alarm and switched the sprinkler system on. To Sushil Dubey, another plant operator on duty that night, the gas seemed like an 'unending, thick mist'.

The water sprinkler system had not been built to neutralize a disaster of this size and the water was falling about twenty feet short of the top of the stack from which the gas was pouring out at a height of around 120 feet. As the intensity of the flow increased, Dey announced the leak over the public address system and the factory's lone fire engine arrived on the scene. That, too, failed to pump water high enough. Workers on the night shift at the sprawling industrial complex began to flee.

The twinkling lights on the company's technological showpiece began to dim as the fumes swirled around it and flowed to the city. The cloud heightened the almost surreal aspect of the plant, an organized tangle of stainless steel pipes and sleek cylindrical stacks, set on the northern edge of the city with farmlands behind it and poverty-stricken hovels opposite.

Slowly, the people of Bhopal in India's Hindi-speaking heartland, began to awaken to horror and death.

Hindus and Muslims, untouchables and high-caste Brahmins, slum-dwellers and affluent traders: the cloud of death spared no one. But the worst-affected were people in the poorer areas neighbouring the plant. The majority of them were malnourished, without proper shelter or sanitation. They were labourers who survived on the wages they got by renting out their crude, wooden pushcarts; they sold *pakoras* and *pethas,* the home-cooked fried potato snacks and sugary sweets of northern and central India; they were many construction workers too.

Slowly, and agonizingly, they began to die, many in their sleep. The city began to cough, to choke and heave, as tens of thousands woke to a suffocating, acrid white-yellow mist that swirled in through open windows, and crept under doors into their bedrooms and kitchens. Then the panic began as people saw husbands, wives, parents and children struck down — gasping for breath, clutching at burning, hurting eyes and chests, frothing at the mouth, as in fits of madness, and then choking on their own vomit and blood. The people of Bhopal began flooding the roads of the town as they tried to flee, picking up children and whatever clothes and cash came to hand, covering their eyes and staggering through a nightmarish world of thick mist, tortured vision and laboured breath.

Seyyid Khan woke at 1.30 a.m. to the shuffling of panicked feet, a stinging in his eyes and coughs of distress around him. Khan lived with his wife and four daughters in Jayaprakash Nagar, a place of rubbish-lined dirt lanes. Khan's home was similar to others there. It had a mud floor, wooden slatted walls and a patchwork roof of corrugated iron sheets, canvas and thick cardboard, weighed down with heavy rocks to keep it from flying off during the occasional high velocity monsoon winds that hit Bhopal. A single 60-watt bulb lit the place.

The colony smelled faintly of urine and failed dreams. But it was home to more than 4,000 men, women and children, the proud possessors of land deeds known as *pattas*.

These *pattas* were handed them earlier that year by the State Government, which conveniently overlooked the fact that they were illegal squatters. What mattered was their importance as a vote bank during future elections.

In his one-room hovel, Seyyid Khan thought at first, as he groggily listened to his wife coughing and wiped his smarting eyes, that a lunatic was roasting chillies in the December night. 'Which cunt of which mother is burning chillies at this hour', he snarled, as he lunged to the door and opened it. But there was no burning chilli pile outside and no madman either. The only thing was that the whole colony, why, the whole world, seemed to have gone mad. There was a stinging whitish mist outside that made him retch and clutch his throat in agony. It felt as if someone had slid a knife down his gullet.

As he gasped and leaned on the wall for support, he saw his neighbours running and screaming. *'Zahreli gas phut gayee, hum sab mar jayenge'* (the poison gas has burst, we are all going to die), he heard someone shout.

The squat and powerfully-built truck cleaner turned to look at his wife. Her head slumped and her coughing stopped. He shook her, trying to revive her. She was already dead. In his fear, Khan forgot about his children and ran into the night, joining a frenzied crush of tens of thousands of people fleeing the fumes on every kind of vehicle imaginable: bullock carts, cycle-and scooter-rickshaws, buses, cars, motorcycles and bicycles. Most of them were on foot.

The traffic jammed, cars broke down, scores climbed onto passing vehicles, others were crushed in the stampede. Mothers were separated from children and husbands from wives as the desperate exodus continued in the darkness of the chilly pre-dawn hours.

The problem was that instead of running against the wind, people ran with it, and the gas, carried in their direction by the breeze, soon overwhelmed them.

The next day, Khan returned home. He had located two of his daughters at a hospital. The others were dead. Khan's employer accompanied him to the empty home, worried that his employee might commit suicide in a fit of depression. 'I couldn't think of anything', Khan recalls. 'All I wanted to do was to run as far as I could. I left them on the bed and they never woke up'. Tears came to him easily as he looked around his home. They flowed even more when he learned that thieves had broken into his hut and stolen his wife's jewellery and what little cash he had left behind. Worst of all, they had taken his *patta,* robbing him of his land and hopes.

Raj Kumar Keswani, thirty-four, balding and bulky, felt he was suffocating. Keswani, a freelance journalist, slept next to an open window in his first-floor bedroom in the family home in old Bhopal. He had been writing until about midnight before going to sleep and was woken up by the noise of people on the roads and the sense of suffocation. At first, Keswani thought the irritating sensation he felt was the onset of the sore throat that accompanied his frequent colds. But the noise alerted him to possible external danger.

Looking down from the window, Keswani was stunned by the bizarre sight of the street, normally empty at this hour, crowded with people running, coughing and crying out, *'Allah, hum to mar gaye'* (Allah, we are dying). Keswani was alarmed and mystified. Obviously, there was great danger but from where and whom? He made two phone calls.

The first was to police headquarters. A policeman, racked by cough, picked up the phone. 'What's happening?' Keswani asked. 'The Union Carbide plant has exploded and there is poison gas everywhere', said the rasping voice at the other end.

The second phone call was to N.K. Singh, staff correspondent of the *Indian Express,* who worked closely with Keswani, and lived in a government-allotted bungalow, about four kilometres away, in the new township. Singh said he knew nothing of a disturbance in the city although the first explanation that flashed across his mind was the possibility of a Hindu-Muslim riot. But he promised to call Keswani back. Keswani did occasional pieces for *Jansatta,* the Hindi daily published by the *Express* chain, the country's largest and most influential newspaper group.

After calling Singh, Keswani realized that his old parents and brother and sister were in the neighbouring room. There had been no sound from them. Checking, he found all of them breathing with difficulty, coughing and rubbing streaming eyes. It was clear that they must all leave the house quickly. As they prepared to flee, the phone rang. It was Singh on the line. The rush of panicky people was then passing his home and he had called out to a group, asking about the exodus. The Carbide plant, he was told, had exploded. 'Get out of there, just leave', Singh snapped at Keswani.

As he talked, Singh watched the great concourse of people

rushing past his house. He and other writers and journalists lived in the 45 Bungalows colony in large, spacious homesteads, paying nominal rents to their landlord, the Madhya Pradesh Government. What amazed him was the orderly procession of the crowds outside his window. People were walking, he thought, as if in a trance. Obviously, they were tired, especially if they had rushed from Jayaprakash Nagar and its surrounding areas, up the winding roads, across the bridges and hills that separated old Bhopal from the new city.

Then his wife began to cough. The journalist quickly placed her and their two children on his scooter and dashed to Arera Colony, where prosperous professionals, including Union Carbide managers, businessmen and powerful government officials lived. The colony is located on a hill, a good ten kilometres from the plant.

At the other end of town, Keswani was on his scooter with his wife and sister. His younger brother, Shashi Kumar, drove the family's other scooter toward Arera Colony. But instead of following Kumar, Keswani drove towards the Union Carbide plant. The mist had begun to clear but the stream of refugees continued. They were coughing and collapsing, rising and walking on, supported by friends and relatives. The women behind Keswani on the scooter were frightened but silent because the journalist had instructed them to save their energy and take shallow breaths. A half-kilometre from the plant, Sunita Keswani told her husband that he was taking too much of a risk and demanded that he go elsewhere.

Her words jolted Keswani out of the rage and despair that had overwhelmed him since he had first been alerted about the tragedy. For the first time he realized he was putting others at risk. Without a word, Keswani turned the scooter around and headed for the home of a friend in the Idgah area, located in the old city, but on high ground. Dropping the women off he sped to Hamidia hospital, as the main government hospital was known. Painted a dirty cream and pink, its formal name was the Gandhi Memorial Hospital. But to everyone in Bhopal, it was just the Hamidia.

The Hamidia was already swamped by a mass of patients. Junior doctors, interns, nurses and senior hospital staff were caught up in an often losing battle to save lives and treat an unknown killer. In the first six hours alone, an estimated 20,000 patients flooded the hospital, sorely testing its facilities for it only had an installed capacity of 760 beds and was not particularly well equipped.

As Keswani arrived, still in his pyjamas, scores of corpses had begun to flow in. Later in the morning, they were to spill out of the morgue on to the roads and lawns outside the hospital's four-storey administrative block—hundreds of anonymous bundles, covered by sheets of white cloth, their exposed faces drawn in the rigours of a painful death, spittle on their lips, limbs contorted.

The sight of the bodies kindled a new anger in Keswani, and a surge of hopelessness. The portly Sindhi, born and raised in Bhopal, after his parents had left their ancestral village of Sukkur in Sind, Pakistan, during the 1947 partition of the subcontinent had, for more than two years, warned of the dangers from the Union Carbide plant.

In three articles, published in September and October 1982, Keswani had predicted the death of his city. The first of the articles in *Rapat,* his own magazine, in Hindi was titled 'Save, Please Save this City'. This was followed by 'Bhopal in the Mouth of a Volcano' and 'If You Don't Understand, You Will Be Wiped Out'. Five days after the second article, a gas leak from the plant started an exodus from the shanties around the factory. Warning sirens rang but the leak was quickly plugged.

Embarrassingly for the company, it took place during a 'Safety Week' when senior executives from corporate headquarters at Danbury, Connecticut, and the Asian head office at Hong Kong were in town.

Keswani's final warning came on 16 June 1984 in *Jansatta,* the *Indian Express* publication, when he repeated his old charges. The next time the factory leaked its deadly gases, he asserted, 'there will not even be a solitary witness to testify to what took place'.

The journalist had begun taking an interest in Carbide's affairs after a friend of his, a plant employee, spoke to him of safety hazards there. But Keswani moved from academic interest to active involvement after a December 1981 incident when a plant worker named Ashraf Mohammed was drenched with liquid phosgene while cleaning a pipe. Khan died the next day. Phosgene is the killer gas that the German Army used in the First World War. Union Carbide uses it to manufacture MIC. Keswani was provoked into investigating the plant. Nine months later, his first piece appeared in *Rapat*.

*Rapat* earned neither a reputation nor money for its owner. No one seemed to pay heed to what were regarded as the ramblings of an obsessed man. Reluctantly, Keswani closed his paper and travelled to the nearby city of Indore where he worked with a local Hindi newspaper. A year later, he returned to his hometown as a writer for *Jansatta*.

On 3 December 1984, as he watched corpses and victims being unloaded at the Hamidia hospital, Keswani came close to weeping. It was a moment of shame, he says.

Yet, there was one incident that gave him some comfort. As he had driven toward the plant early in the morning, the reporter had watched a burkha-clad woman, a child in her arms, running, then walking and faltering in her flight. As she tired, a car passed her by. The woman stopped the car, pushed her child into it, and watched as it drove off. There were no questions asked, no answers needed.

Elsewhere in the city, too, bonded by a common courage in the face of an unknown fear, strangers helped each other to get to hospitals, to reunite families split by the initial panic, to transport the sick and needy to safe places. In the initial hours, as state and city administrators still gathered their wits, information and officials, a massive voluntary public effort saved the lives of thousands afflicted by the gas.

A guard roused Arjun Singh, the state's Chief Minister, as he slept in his sprawling official home. Located on Shamla Hill, with a commanding view of the Upper Lake that fed the city's water system, the place had been home to Arjun Singh since 1980 when he was picked to lead the state by Sanjay Gandhi, the late Prime Minister Indira Gandhi's younger son. Through the skilful use of power, patronage and pressure, Singh (whose father had resigned from politics on charges of corruption) had quickly become one of the most influential Chief Ministers in India. The events of this night were to catapult him firmly into national and international prominence (Singh became Governor of Punjab the next year and several months later, national Minister for Communications).

The guard called Arjun Singh on the internal phone saying

'something bad' had happened in the city. By then, the Chief Minister was feeling the sting in the air. It reminded him of being teargassed as an opposition legislator in the late 1970s. Instinctively, he washed his face.

He tried to call the District Magistrate, Moti Singh, and Swarj Puri, the city's Police Superintendent. Neither official was at home. He left word for them to call him back. About 3.30 a.m., Arjun Singh had his first visitor, Rewanath Choubey, the Health Minister, who explained to him that there had been a leak from the Union Carbide plant. The phones began to jangle and buzz with reports of the disaster.

For a time, Singh and his aides were so stunned and shaken that they did not know how to cope with the emergency. 'We were quite nonplussed', he says. But at 5 a.m., when the police announced that the gas leak had ended and urged people to return home, Singh drove to the ravaged areas, accompanied by the Health Miniser.

While the Chief Minister was racked with indecision, Puri, the police chief, had driven through the cloud of gas at about 1.30 a.m., after a constable had run to his home and alerted him to the *gharbar* or commotion in the old city. Like N.K. Singh, the journalist, Puri's first thought was that a sectarian riot had erupted. He was still unsure of the cause of the problem when he saw people streaming up the slope past his home.

A strapping, affable man, with a bushy moustache, Puri is one of Salman Rushdie's *Midnight's Children* in real life. He was born three hours after India became an independent nation on 15 August 1947. His mother named him Swarj or freedom. The handlebar moustache was a result of his years in the army.

As Puri's car rushed to the police station in the old city, it met the gas head on. The driver panicked. Coughing and choking, the policeman pleaded with his chief to change direction. But Puri ordered him to go on. By the time he reached the control room, Puri was feeling the strain on his eyes and lungs.

The control room—a rather grand appellation for the small room with paint peeling from its walls, scruffy chairs and tables—was filled with policemen breathing with difficulty and vomiting.

Meanwhile, the army had been alerted by a retired brigadier

who ran Straw Products, a factory located near Union Carbide. Brigadier M.L. Garg asked the army's area sub-commander for help saying he needed to evacuate the factory. He got the assistance. By about 3 a.m., the army, accustomed to handling natural disasters, was fully operational. It was among the few organizations that functioned effectively that night with soldiers, in a swiftly-moving fleet of trucks from the Electrical and Mechanical Engineers Centre, fanning out to search for victims. Puri ordered his policemen to assist in the search. The troops later also kept vultures and stray dogs from getting at the dead.

At police headquarters, Puri met H.L. Prajapathi, the slightly-built Additional District Magistrate. Prajapathi said that he had spoken to Jagannath Mukund, the factory's works manager, at his home and given him news of the disaster. Mukund's first reaction, Prajapathi asserted, was one of incredulity. How could it have happened? Mukund had asked. The plant was shut down, he had told the official.

Before ordering the Carbide manager to go to the plant, Prajapathi had asked him about medical treatment. The way to help people, he was told, was to splash water in the eyes and wipe the face and mouth with a wet cloth. Prajapathi recalled that Mukund had said, 'it is not known to kill'.

Puri made three calls to the plant but could not get through. He called Mukund's home and learned that the factory manager had just left for the plant. In anger and frustration, the police chief ordered A.K. Singh, a young police inspector, to rush to the Carbide plant and find out what was happening there. Singh arrived about the same time as other district officials. Mukund was also there, having driven there after talking to Prajapathi. He had sent his security officer to the police station to inform Puri that the leak had been stopped.

The leak had not been stopped. What had actually happened was that Tank No. 610 had simply disgorged most of its load of forty-three tons of methyl isocyanate into the night sky.

Back at the station, Puri asked the plant's security officer to identify the gas and its antidote. The man said he didn't know. Puri ordered him out of the police station.

By then, the air had cleared. But the effects of the gas could still be seen. Human and animal corpses were scattered in the gutters and lanes of the city. A mother and her child clutched each other as they sprawled on Hamidia road, one of the main thoroughfares, united in death as at birth.

What was the gas that had escaped? asked police officials at the plant. Methyl isocyanate, said K.V. Shetty, the night shift superintendent, who had been at the place throughout the incident. None of the law enforcement people had heard of it. Mukund later spelled the name out for them.

Among the worst-hit areas were the railway colony and the railway station. At least seventy-three persons died in the railway workers colony as the gas rolled in. Those at the railway station became the next victims. Porters, passengers and railway officials were soon crying, rolling about and vomiting on the platforms. The control room at the station was soon a mess of vomit and excreta as staff collapsed at their posts.

H.S. Bhurve, the station superintendent, rushed out and waved on an incoming train—its windows shuttered against the winter cold—saving hundreds of lives. Bhurve then went to the control section and sent desperate messages to railway stations around Bhopal to stop trains from travelling to the city. He was later found dead near his office.

The devastation was now almost complete. The hospitals were overflowing with patients and unable to cope with the rush. The entire population of old Bhopal (estimated at more than 200,000), barring a few exceptions, had fled to other parts of the city and outside. Some travelled to Sehore, Gwalior, Indore and countless other towns and villages. The old city, which owes its origins to Raja Bhoj in the twelfth century, and an Afghan adventurer in the eighteenth century, lay deserted. Everyone who could flee had done so, leaving their homes unlocked and open.

Rescue teams of troops, police, local citizens—including well-known thugs—and voluntary organizations such as Mother Teresa's Missionaries of Charity went into the abandoned homes and hovels, dragging out corpses and the injured. They flung the living onto stretchers and vehicles which went scudding to the city's hospitals and clinics. The dead were piled onto trucks which went to the main morgue at the Hamidia or straight to the Muslim graveyards at Jehangirabad and the Hindu cremation site at Cholla.

At the funeral pyres at Cholla, members of the right-wing Rashtriya Swayamsevak Sangh, in their brown shorts and white shirts, placed the bodies on long piles of heaped wood donated by Hindu contractors. With a full bearded priest chanting the final *mantras,* RSS workers arranged flowers and poured kerosene on the bodies, gently uncovering the white shrouds from the faces of the dead, before torching the mass pyre. Many of the dead were unknown and were cremated without identification. Some had just a number written by a morgue clerk on the cloth that covered them. An RSS organizer maintained a register of the bodies.

As the funeral pyres burned, another 100-yard-long line of wood was kept ready for the scores of fresh bodies pouring in. Trucks disgorged corpses every three or four minutes. Much of the grief was silent as relatives huddled in shocked clusters to look at the dead. But there were those who could not hold back their sorrow. *'Is seh accha tha ke main mar jaon'* (It is better that I should have died) cried a father as he lifted the sheet covering the face of his five-year-old daughter. He collapsed and was led away by relatives. RSS workers and family members placed the girl in a shallow grave and covered her with dirt. Another child was also lying in the pit, feet pointing towards the head of the girl, his eyes open and unseeing. Before he too was buried, photographers took his picture. That photograph (reproduced on the cover of this book) would soon come to represent the horror of the tragedy to the world. As the cremation fires burned, weeping relatives, their eyes bandaged with cloth—the gas had scarred the outer tissues of the eye—moved from corpse to corpse in the lines awaiting cremation, looking for parents and children, husbands and wives.

In Jehangirabad, the crush of bodies was such that the Muslim caretakers of the graveyard had to open up old graves and bury many in common graves holding eleven bodies or more. Here, as elsewhere in the city, local committees sprang up to organize the burials. There was no official assistance.

Later in the day, the city corporation, never known for its efficiency, began sending out trucks to remove the bloated carcasses of buffalos and cows which were lying everywhere. But the trucks just didn't have the equipment to lasso the stinking carcasses and lift them. Two days after the disaster, cranes provided by the army and Bharat Heavy Electricals Limited, the

Central Government's power plant manufacturing firm, were still removing dead animals, tossing them into trucks and dumping them at Nishad Pura, about five kilometres north of the city.

Here, at the traditional animal scavenging and flaying centre, the carcasses were sprayed with lime and salt and buried in large graves amid an overwhelming stench of rotting flesh. Vultures circled lazily in the sky above or sat near the site, often jumping in to tear at a carcass.

Meanwhile, Arjun Singh had begun to get his act together. He had visited the hospital and was horrified by the number of people requiring medical attention. His special secretary, Krishnan, returned from discussions with Union Carbide's factory managers and pointed out that the company now said there was no chance of another leak.

Singh then ordered the arrest of the works manager, Mukund, along with four other UCIL employees and announced the closure of the factory.

His Government sent out appeals for medical help and drugs to surrounding cities in the state. Messages began to flow between Bhopal and Delhi. Singh sent word to Prime Minister Rajiv Gandhi of the accident.

The first that the country heard of the disaster was on the 8 a.m. newscast on the third of December over the government-run All India Radio. The broadcast did not name Union Carbide but the description of the company at which the disaster had occurred was horribly clear to Natarajan Kumaraswami, the firm's manager in Delhi, who dashed immediately to his office. Kumaraswami, a twenty-year employee at Union Carbide handled company applications for licences and liaison work with the Central Government. He spent the rest of the day fielding questions from Indian officials, talking with Mukund in Bhopal and communicating with managing director Vijay Gokhale in Bombay. Later that night, he was called by R. Natarajan, a director of the company who was at that time at the multinational's corporate headquarters in Danbury, Connecticut. An emergency meeting of corporate executives there had picked Natarajan, who operated out of the company's Asian headquarters in Hong Kong, as a coordinator from the American side.

In Bombay, on the morning of the third, Vijay Gokhale, the

managing director of the Indian subsidiary, Union Carbide India Limited, summoned an emergency meeting of his top executives in his ninth-floor office. Gokhale briefed them on the information he had received, which was quite scanty, and then identified two priority areas.

One was to gear up for the new responsibilities which the disaster would involve: thus, Gokhale took upon himself the responsibility of handling all aspects of the disaster with the Government and the travelling it would involve. He caught the flight to Bhopal that evening. K.S. Kamdar, the vice-president of the Indian subsidiary's Agro-Chemicals Division, was placed in charge of the Bombay office to free the managing director from daily chores.

Secondly, the crisis management team decided that it was critical for the office to maintain regular contact with Bhopal. Efforts to persuade the Bombay Telephone Department to open a hot line to Bhopal failed and the company had to content itself with booking lightning calls at regular intervals. Of these, only two or three went through.

Gokhale was not the only person having problems communicating with Bhopal. In New Delhi, Nilay Choudhury, the fifty-one-year-old head of the country's Water and Air Pollution Control Board, informed about the disaster only that afternoon, tried to call the Madhya Pradesh Pollution Board, without luck. He made a mental note to tell his wife that a trip to Agra, to celebrate their sixteenth wedding anniversary the next day, was off.

Choudhury called Kumaraswami, the company representative in Delhi. Everything was under control, he was told. The UCIL official outlined what could have happened. Choudhury summoned Professor J.M. Dave of Jawaharlal Nehru University's School for Environmental Sciences, and asked him to get as much material as possible on methyl isocyanate.

As Dave set about collecting data, T.N. Khushoo, the Environment Secretary, called. Did Choudhury have any details on the accident? he asked, reflecting the lack of information in New Delhi, as much as in Bhopal, about the disaster. Adding to the confusion was a call from the pollution board at Bhopal saying there was panic in the city and experts were needed. Dave was soon back with details on the chemical and Choudhury decided on the tests that his staff must carry out to confirm whether Bhopal's air and water were safe.

Assembling a team of six specialists, Choudhury asked them to prepare kits to check the levels of hydrogen cyanide and urea in the air and to look for cyanide and cyanate in water. Early on 5 December the experts flew to Bhopal, joining the army of other specialists and scientists that had descended on the city.

The tests found the air and water safe in Bhopal. They confirmed a statement the Prime Minister, Rajiv Gandhi, had made immediately after the disaster that both elements were uncontaminated. However, Choudhury says that Gandhi made his statement without an official study and that it was only meant to reassure the people of Bhopal. Interestingly, three days after the gas leak, the Delhi team's experiments turned up traces of cyanide in the air near the MIC storage area, where the leak had originated.

Among the non-technicians at the site of the disaster was a team from the Central Bureau of Investigation which had arrived from New Delhi on the day after the incident. The group was led by Brij Shukla, a cigar-smoking, gruff, CBI veteran, who immediately took control of the plant. Shukla put a ban on visits to the MIC area by company employees, seized factory records and began interrogating UCIL employees.

In the meantime, the Indian Government set up a seven-member crisis cell to coordinate the work that needed to be done. A hot line linked Delhi and Bhopal. The special team worked out of a large room in the Rashtrapati Bhavan, the vast presidential residence designed by Edwin Lutyens for the British Viceroys who once ruled India. The room belonged to the Cabinet Secretariat, a government branch that supervises the functioning of India's main intelligence operations.

The secret panel including Choudhury and members of the Ministries of Health, External Affairs, Home Affairs, Chemicals, Environment, the Cabinet Secretariat and a representative of the Council for Scientific and Industrial Research. Every day, the group met and received status reports of the situation in Bhopal and assessed what needed to be done. Its reports went to the Prime Minister.

Rajiv Gandhi, India's youngest Prime Minister, had been

through the most traumatic year of his life. His mother, a seemingly indestructible woman who had towered over the country as its chief elected official and mother deity for nearly twenty years, had been murdered by two Sikh guards at her residence in October. Her unexpected death thrust upon her elder son, a forty-year-old airline pilot and political novice (his political experience was limited to the revamping of the party in his capacity as its general secretary and four years as a Member of Parliament), the multifarious duties and responsibilities of the prime ministership. He was formally sworn in by President Zail Singh and immediately confronted India's biggest political crisis since independence as across northern India Hindus and Muslims slaughtered Sikhs to avenge the assassination of Mrs. Gandhi. The new Prime Minister tackled the issue firmly, if slowly. The danger passed and the country prepared for fresh general elections.

Barely a month after the tragedy at his home at 1, Safdarjung Road, came Bhopal.

Gandhi interrupted his election campaign and flew to the city, announcing a Rs. four million relief fund for the victims. The Prime Minister stayed in Bhopal on 4 December for a little more than three hours. He visited the Hamidia hospital where patients were sharing beds and sleeping on the floor. Gandhi was visibly moved by the suffering there, especially in the children's ward, where infants, with parents and doctors watching helplessly, screamed in agony and thrashed on the beds and tables as their battered lungs tried to suck in air.

During his visit, Gandhi declared he would ask Union Carbide to pay compensation. He also said that his Government would review procedures on the siting of hazardous industrial units and existing safeguards.

Soon after Gandhi left, Gokhale, the UCIL chief, met Arjun Singh, the Chief Minister. Singh was polite but non-commital. Gokhale was the supplicant, profusely apologetic. The company official offered an *ad hoc* payment of Rs 15,000-20,000 for every death. This, he said, would not be connected to any eventual compensation. Singh said he would give Gokhale an answer the next day.

On 5 December, Gokhale went to the plant early in the morning to meet Mukund and the others who had been arrested. They discussed possible causes of the accident and its horrifying aftermath. Gokhale assured the shaken men that he would take care of their interests. Later in the morning, Gokhale drove towards the Chief Minister's home to keep the appointment on the compensation issue. Halfway there, a company car flagged him down with a message. The Chief Minister, he was told, wanted him to go to the plant because Vasant Sathe, the country's Chemicals Minister, was already there.

Gokhale returned to the plant, escorted Sathe around the area, answered his questions and added that the Indian subsidiary, not the corporate giant, was offering a sum as interim relief. Sathe reacted to this at a press conference just before returning to Delhi. Compensation to the victims, the minister declared, must be 'at American levels'.

Arjun Singh did not give Gokhale another appointment, although the company executive did request a meeting for the corporation's chairman, Warren Anderson, who was on his way to Bhopal.[2] Gokhale was earlier assured of both police protection and a meeting for his chief.

On 6 December, Gokhale returned to Bombay to receive Anderson, the sixty-three-year-old, white-haired son of working-class immigrant parents, who presided over the fortunes of one of the world's largest corporations.

In more ways than one, Warren Anderson symbolized the American dream. The son of a Swedish carpenter who had emigrated to the United States, the young Anderson worked with his father and delivered newspapers to supplement the family income.[3] He won scholarships to college in football and math, served with the Navy in the Second World War (but never saw combat) and then, when the fighting ceased, joined Union Carbide. He started as a lowly salesman and rose steadily up the corporate ladder—general sales manager in New York, vice president of the international division, president of the profitable chemicals and plastics group—until he became president and chief executive in 1977. The company he ran spanned 700 units in over thirty nations. Anderson earned nearly a million dollars a year in salary and perks. He was low-profile, enormously capable and an efficient but quiet leader of his troops.

Anderson had visited India once before, in 1978, during a familiarization trip when he took over as top executive. At the time, Vijay Gokhale was a manager at the Calcutta battery plant. But this was Anderson's first trip to Bhopal.

The head of Union Carbide had heard of the disaster in Washington on 3 December, just before he was to return to Danbury after attending a function at the Kennedy Arts Centre. Alec Flamm, one of his vice presidents, had telephoned to tell him of the incident. Anderson had a bad cold and went straight to his home in Greenwich, Connecticut, after returning from the capital. There he listened to radio updates of the disaster, watched television descriptions of conditions in Bhopal. Like his subordinate officers in Delhi and Bombay, he spent a lot of time on the telephone discussing what the company and he should do.

Anderson decided to travel to Bhopal. He ordered a technical team also to visit Bhopal, assess the reasons for the disaster and help in any way possible. He arrived in Bombay on 6 December jetlagged and tired. Keshub Mahindra, the UCIL chairman, who belonged to one of India's top industrial families, Gokhale and Anderson discussed the immediate priority—compensation.

The next morning, the three men boarded the first flight to Bhopal. Waiting for them at the city airport was Swarj Puri, the Police Superintendent. Puri had sent a wireless message to the pilot saying that the three men should alight first. He received the businessmen and drove them to the Union Carbide guest house, located above the Chief Minister's residence.

The high steel gates were shut and police guards kept a horde of news reporters and camera crews, local and international, at bay. A short while later, the District Magistrate, Moti Singh, arrived, dodging a barrage of questions and microphones. 'You are under arrest', Singh told the executives. They were being charged, he said, with 'criminal negligence and conspiracy'.

The decision to arrest Anderson and the Indian executives of the company was apparently Arjun Singh's but it is still not clear whether he had Delhi's concurrence or clearance. What might have prompted the decision is the fact that elections to Parliament were on and Singh's hold on his state was being put to the test. The arrest of the head of the multinational involved in the Bhopal disaster could have been viewed as enhancing his authority. A senior Central Government official says he was as surprised by the arrests as anyone else.

In the event, the arrest was resented by American Government officials and Indian business leaders. A White House spokesman objected, saying Anderson had been assured safe passage by Indian officials. The American diplomat who was accompanying the industry leader from Bombay was not allowed to meet him at the guest house. In the afternoon, Indian journalists like Ashtesham Qureshy of the *Hindustan Times* wanted Sudeep Banerjee, the State Government's spokesman, to distribute sweets to celebrate the arrrest. Banerjee, a cheerful bureaucrat, who wears loose-fitting suits, issued instead a statement by the Chief Minister announcing the arrests. Anderson and the others, Arjun Singh declared, had 'constructive and criminal liability for the events that have led to the great tragedy'. The Government, the Chief Minister announced, 'cannot remain a helpless spectator-....and knows its duty toward the thousands of innocent citizens whose lives have been so rudely and traumatically affected'.

There were seven charges listed against the men: criminal conspiracy and culpable homicide not amounting to murder were the main ones. The others included causing death by negligence, mischief in the killing of livestock, making the atmosphere noxious to health and negligent conduct in respect of poisonous substances.

A brief, small demonstration calling on the Government to hang Anderson took place ouside the gates of the Carbide property, which also includes a research and development centre and administrative offices. The rally dispersed quietly.

Outside the locked gates of the guest house, correspondents of the *New York Times,* the *Washington Post* and the *Los Angeles Times,* among others, weary of waiting for news began typing drafts of their stories, leaning on the hoods of taxis and jeeps.

A sudden commotion was caused by the exit of four men, three Americans and an Indian. The Americans jumped into waiting cars outside the gate and dashed off. Was one of them Anderson? the journalists wanted to know. No, they were told, it was the Carbide technical team which had preceded him. But the Indian from Union Carbide hesitated, as he looked around for his car.

Just as he spotted the vehicle and entered it, a reporter blocked the road and said: 'Dr. Awasia, we have been wanting to talk to you for a long time'.

'I'm not Dr. Awasia', huffed the little man in spectacles.

'Dr. Awasia, your accent gives you away', said the *New York Times* reporter.

'You're trying to hijack me, I won't be railroaded into talking like this,' yelled the man. Bipin Awasia was Carbide's health director at Institute, West Virginia, where the only other plant making MIC by methods used at Bhopal was located.

'Dr. Awasia', the reporter continued, as other journalists suddenly clustered around the car, sensing a prospective victim, 'Your plant has leaked a gas that has killed hundreds of people, the doctors here don't know what to do, I don't think you can afford to refuse to talk to us.'

'All right, I will talk with you, but not here, we will hold a press conference very soon, but right now I have to locate the medicines I have brought,' the Indian-born doctor added, his annoyance suddenly departing as he realized he was cornered. (At the point, he wasn't material to the real story, the arrest of Anderson, so the reporters let the rattled man go).

A different drama was being played out inside the guest house. Word had come from the Chief Minister's office that Anderson was to be released. It had to be done quietly. Anderson was freed on bail, surety of which was set at Rs. 20,000, and a personal guarantee that he would leave the country soon. To avoid the crowd outside, Puri took the American out through an old disused side entrance to the airport. By the time the reporters realized what was happening, Anderson was on his way to Delhi in a state government plane.

Responding to criticism from opposition leaders and local reporters on the six-hour drama, Banerjee, the government spokesman, said that there had never been any intention to prosecute Anderson. 'He is not required in the investigation', Banerjee snapped. 'His presence is not desirable and might provoke strong passion against him.'

Protected from the press by American Embassy officials, Anderson was rushed into a meeting with the U.S. *chargé d'affaires*, Gordon Streeb. He met the Indian Foreign Secretary,

Maharaj Krishen Rasgotra, for what was officially described as a 'long and friendly discussion' the next day before returning to the United States in a company plane.

Upon his return, he called a news conference at Danbury to outline his thoughts. 'The name of the game is not to nail me to the wall, but to provide for the people', he declared. He denied charges of criminal liability, saying the Bhopal plant was as good as the main factory at Institute, Virginia. But the company, he acknowledged, had 'a moral responsibility and we are not ducking it.'

Anderson, knowingly or otherwise, had been incorrect in his statement about the Bhopal plant. So had Warren Wommer, a former works manager at Bhopal, who had told a group of reporters at the UCIL administrative offices on 5 December that there was 'no difference' between safety procedures at Institute and Bhopal. Both men asserted that the back-up systems at both the plants were the same. 'We follow the same safety precautions in our plants in the US as in other countries', Wommer said. This was not correct. The systems at Institute were computerized and state-of-the-art, which they weren't at Bhopal.

It was also true that Union Carbide had a long history of industrial accidents despite the precautions it claimed to be taking. In April 1986, Carbide was fined $1.4 million by the US Occupational Safety and Health Administration (OSHA) for 221 violations of fifty-five federal health and safely laws. The violations, OSHA said, took place at the Institute, MIC-based, pesticide plant. 'We were just surprised to find constant, wilful, overt violations on such a widespread basis', said US Labour Secretary William Brock.[4] The installation was upgraded at a cost of $4 million after the Bhopal tragedy. However, despite the improvements, the plant continued to be dangerous. Ironically, Jackson Browning, a company vice-president, had first gone on record saying a Bhopal-type situation couldn't take place at Institute. 'We can confidently say, "It can't happen here,"' Browning said.[5]

In August 1985, less than six months after Browning's statement, the plant leaked poisonous aldicarb oxime and methylene chloride gases that sent more than 100 people to local hospitals.[6] Carbide officials said at first that the gases that had leaked were less toxic than methyl isocyanate. But Awasia, the plant physician, admitted later that, 'very little epidemilogical work has been done on the substance.'[7] The company also erred otherwise. The leak had continued for twenty minutes before the plant alerted local authorities. The public warning siren was sounded only after town officials had been informed. Said a harried Robert Kennedy, then president of Carbide's chemicals and plastics division (and now chairman): 'We don't know precisely when to trigger an emergency response'.[8]

Union Carbide's experience of industrial disaster goes back to 1930 when workers at a subsidiary known as The New Kanawah Power Company building a hydro-electric tunnel in West Virginia, near the town of Gauley Bridge,[9] were affected by the work they were doing. Reporters say that the company knew (but withheld the information) that the workers were cutting through silica. Dust from silica causes silicosis, a lung disease that can be fatal. The results were predictable; a total of 476 workers eventually died of silicosis.[10] The incident has been described as the century's worst industrial tragedy in the United States.

Union Carbide has also tested one of its best-selling pesticides, Aldicarb, on human volunteers in Panama. It reported its findings in an unpublished document submitted to the World Health Organization.[11] In 1976, vinyl chloride workers at a Carbide plant in South Charleston accounted for six cases out of a worldwide total of sixty-three of a rare cancer associated with the chemical. Workers filed for compensation against the company.[12]

Carbide uses methyl isocyanate as well as several other compounds, including aldicarb oxime, to make Temik. Temik has been found in groundwater in several American states.[13]

The corporation drew praise for its cooperation with local authorities in responding to and resolving that crisis. In 1978, Union Carbide withdrew a catalyst used in the manufacture of polyurethane foam from the American market after more than 100 workers handling the substance began experiencing bladder paralysis.[14]

In 1979, Carbide's health director at Jakarta, Indonesia, quit her post over wilful violations by the company of safety procedures.

The official noted that workers were suffering from kidney disease and breathing problems, and found traces of mercury in the well water that the workers drank. The following year, inspectors on the battery inspection line began developing behavioural problems and six were removed from their jobs.[15]

The accidents don't stop there. In May 1986, nine tons of ethylene oxide, which is highly poisonous and inflammable, spilled from the Institute plant into the Kanawana river. A State official accused the company of waiting for six hours before notifying authorities of the accident. There were no reported casualties.[16]

The next month, Carbide agreed to pay $33.3 million in cash and credit to buyers of industrial gases to settle charges of price-fixing. The consumers had charged in a class action suit that five producers including Carbide had conspired to fix the prices of industrial gases between 1974 and 1980.[17] Carbide's record therefore wasn't exactly satisfactory much as Anderson and Wommer would have liked it to be.

And the Bhopal facility was certainly more poorly maintained than perhaps anything the company owned in the United States.

At the Indian plant, technicians and workers would realize toxic gases were escaping only by the stinging of their eyes for the acrid smell. Most of the factory's instruments were manually operated and often faulty. The inadequate facilities at Bhopal reflected poorly on the reputation of the firm that had done business in India for more than fifty years and was a household name.

Union Carbide was (and continues to be) renowned in India for its Eveready torch batteries. It manufactured nineteen different kinds of torch, transistor and appliance batteries, including specialized ones for defence equipment. The involvement with the military, the company said, was its 'proud contribution' to the country's security. In addition, it produced flashlights, lanterns, arc carbons for cinema projectors, photoengraver's plates and strips for printing, chemicals and pesticides. It owned one of the country's best-run research and development centres on pesticide use. It had a network of fourteen plants spread across the country and was reputed to be a solid investment by those playing the stock market in Bombay. Its managers were competent, it had a productive work force, paid its employees well by Indian standards, and made contributions to charity.

Its sales in 1984 touched Rs.22.8 million,[18] an increase of nearly Rs. 1.27 million over the previous year. But profits in the company's annual report were pegged at Rs. 8.21 million, a Rs. 1 million drop from 1983. Its fixed assets were placed at Rs. 4.67 million. The biggest earner was Eveready batteries, with nearly half of the total sales. One of the major losers was the pesticide plant at Bhopal, which manufactured 1,180 tons of goods, less than one-fourth of its registered licensed capacity of 5,280 tons per year. That production figure was a drop of 320 tons from 1983.

The figures tell an ironic tale. The battery division was Union Carbide India Limited's oldest operation, yet it was its biggest money-spinner. The pesticide plant at Bhopal was the most modern of the company's units in the country, perhaps the most advanced such plant in Asia, and its most consistent loser as well. Company officials say it lost around Rs. 20-40 million every year.

Union Carbide began doing business in India in 1905 but it was not until 1924 that it began a battery assembly factory at Calcutta, putting together components imported from Britain. In 1956, the company was named Union Carbide India Limted.

Union Carbide Corporation controlled the Indian company from Danbury and had a 50.9 per cent share of the equity,[19] unusual for a joint venture corporation in India. Normally, the equity holdings permitted for foreign companies wishing to set up shop in India is restricted to forty per cent under the Government of India's Foreign Exchange Regulation Act (FERA), which was passed by Parliament in 1973. However multinationals with sophisticated technology and export-related subsidiaries were made the exceptions to this rule. Union Carbide was one such and was allowed to maintain an equity holding of just under fifty-one per cent, down from sixty per cent before FERA regulations came into force.

The idea of a pesticide plant came to Carbide in the 1960s, when India was delighted and impressed by the successful 'green revolution' in its north. Modern methods of cultivation and the high use of fertilizers and pesticides had led to a several-fold increase in foodgrain production, especially in the Punjab. Union Carbide saw itself investing in a profitable, growing market and at the same time endearing itself to the Government by doing something that would help the country's drive for self-sufficiency in food.

In 1966, Carbide proposed to the Central Government that the company be allowed to set up a modern pesticide plant. The Government okayed the proposal. The first site chosen for the plant was Trombay, a small township near Bombay. In fact, the company started a pilot plant there but decided unilaterally to move out in 1968. That was when it picked Bhopal and the Kali Parade area as the spot to build, first, a formulation or finishing plant and later a methyl isocyanate manufacturing unit. The MIC would be used to produce Sevin and Temik, brand names for pesticides.

Kali Parade, the site of the proposed plant, was on the outskirts of the fledgling city of Bhopal and seemed an ideal place for an industrial plant. There was little human habitation in the area. Bhopal was still being built and the railway station was nearby. There was plenty of water from the Upper Lake at Bhopal. Labour was cheap and easily available.

Bhopal seemed an excellent choice for other reasons as well. It was linked to the major ports of India by road and rail. Bombay and Calcutta, where the company had major producing facilities and corporate offices, were almost equidistant from the state capital. Delhi was an easy 600 kilometres north, while the port city of Madras was also connected by road and regular trains. Airline services had started and the State Government welcomed the company's investment as a means of developing its virtually non-existent industry as also opportunities of employment.

The plant was started in 1969 on a five-acre plot in a region identified as an industrial area in the Bhopal Master Plan. A small alpha-naphtol plant came up in the Carbide complex and the chemical was mixed with MIC imported from Institute to make Sevin, the pesticide.

In 1970,[20] the UCIL unit applied for and got permission to manufacture methyl isocyanate, a clear, almost colourless liquid with a low evaporation point. MIC is one of the most toxic substances known to man. The approval came through on 31 October 1975, as licence no. C/11/409/75.[21] It was issued by the Ministry of Industry and Civil Supplies in New Delhi. The end product was registered with the Insecticides Board and, company officials said, the actual manufacturing process with the Ministry of Chemicals.

The plant design was done locally, on the basis of designs supplied by the parent firm. Gordon E. Rutzen was one of the engineers involved in the project. Here, and throughout the history of the company, Union Carbide Corporation was always an omnipresent force. The detailed design work was done by Humphreys and Glasgow Consultants Pvt. Ltd., a Bombay-based subsidiary of Humphreys and Glasgow of London.

The aim of the MIC-producing unit was to stop expensive imports. Under an agreement signed with the Indian Government, Union Carbide agreed to transfer the technology involved in manufacturing the pesticide to its Bhopal unit. The Government sanctioned the MIC plant because UCIL had agreed to make the whole operation an indigenous one in five years to save foreign exchange. In 1980, the MIC segment at the Bhopal unit became operational. By 1981, the whole plant sprawled across seventy acres. But it was doomed to be an unprofitable venture.

The worst drought in thirty years hit the country in 1979 and farmers took out government loans to meet the emergency. The loans came due in 1980 and farmers with much less money to spare turned to cheaper pesticides produced by small-scale producers, instead of Union Carbide's Sevin. No one disputed the effectiveness of Sevin on pests which attacked cotton, rice, lentils and vegetables. But the financial crunch forced the farmers to look elsewhere.

The Union Carbide plant's profitability began to dip. Competition from new pesticides—safer, inexpensive synthetic pyrethoids—began to cut into the local market. Although the Bhopal unit made a modest profit in 1981, it was never a truly economic proposition. At the time of the disaster, its owners were planning to dismantle the plant, and send it to potential buyers overseas.

A reason for the unit's economic disability was the shortsightedness of its designers who, in consultation with Union Carbide Corporation, decided to build an expensive $2.5 million alpha-naphthol plant. The result was problems with plant equipment that was not suited to Indian conditions. The plant was shut for modifications in 1978, barely a year after it began production. It failed again in 1983, when it was restarted after an investment of nearly $2 million.

There were more mistakes in the choice of technology. One was the use of phosgene, the killer gas used in World War I by Germany, as an intermediate in the manufacture of MIC. Other companies manufacturing MIC such as Bayer AG, the German chemical giant, do not use phosgene but two non-toxic intermediates as part of the manufacturing process. Another problem was the large amounts of MIC stored in the plant, although the Union Carbide manual itself says a tank as large as No. 610 should never be more than half full. It had a capacity of 15,000 gallons and was nearly three-quarters full at the time of the accident.

As early as 1972, an internal Union Carbide report said, after a study of the Institute plant, that almost every item in the MIC unit there had 'failed and been replaced since start-up'.[22]

'If another facility is built to produce MIC, based on the process used in Institute, materials of construction at least as good as those presently used in the facility at Institute will be required', the report added.[23]

Two key engineers from Union Carbide were involved in the supervision of the Bhopal plant — Gordon Rutzen and John Couvaras of Houston. Couvaras was named UCIL's Bhopal manager in 1972 and spent nine years in India before being reassigned to West Asia. Despite all this the construction of the Bhopal plant was certainly not up to the standards of the company's US installation.

There were other questions about the safety of operating procedures at the plant. A former American managing director of UCIL, Edward Munoz, said in an affidavit that corporate headquarters turned down a UCIL demand that large amounts of MIC should not be stored at Bhopal.[24] Continuing with storage in one-ton containers, the corporation felt, would be uneconomical and inefficient. Munoz said he had argued for the token storage of MIC, preferably in small containers.[25] Carbide tried to dismiss Munoz's testimony in a pre-trial hearing in New York on 24 October 1985, alleging that the man was unreliable. James H. Rehfield, a company vice president, told lawyers examining him that Munoz was charged by the company 'of prostituting employees in order that they maintain their jobs, literally on the job...of being consistently a liar; of cheating in every way, shape, and form in expense accounts, and right on down the line.'[26]

The setting for the disaster was also partly formalized in the mid-1970s. In this period, the state's Town and Country Planning Board classified the Carbide plant as belonging to the general industry category instead of calling it a hazardous unit as it could have done under the city's 1975 Master Plan. Thus, instead of being asked to relocate, UCIL was allowed to stay.[27] Sixteen smaller industries were however located in a new region fifteen miles away, where the wind direction was unlikely to blow pollutants towards the city.[28]

The board overruled M.N. Buch, a former board secretary and administrator of Bhopal, who had told Carbide to shift to the new site. Soon afterwards, Buch was sent to another department after he differed with the then Chief Minister, Shyama Charan Shukla, on a totally unrelated issue. Buch says he had cleared the factory's developmental plans but had not ratified its building projects. Sanction for such construction is granted by the local municipal corporation.

There is no doubt that haphazard town planning and unchecked urban growth contributed to the *magnitude* of the disaster. A look at the population figures immediately makes this apparent. When the Union Carbide plant was begun, Bhopal's population was just above 300,000. By 1981 the national census placed the city's population at 700,000. In 1984 the population was estimated at close to 900,000 with a constant influx of job-seekers from rural areas across the country. Most of them were uneducated and worked as unskilled, casual labourers at construction sites or took on service jobs at the existing housing colonies as dhobis, sweepers and scavengers, cycle repair mechanics, shoeshine boys, owners of tea shops and cigarette kiosks. A large majority of these people were squatters in the areas around the plant. Most of them died when Tank No. 610 leaked.

The causes of the runaway reaction in Tank No. 610, which vaporized or polymerized more than forty tons of MIC, are still disputed by the Indian Government, environmentalists and Union Carbide.

But management culpability is clear.

On the night of the disaster, four out of five major safety systems at Bhopal failed to work due to indifferent maintenance, poor management and bad planning. At the time of the disaster, the MIC plant was shut down for repair. The flare tower, used for burning carbon monoxide, was shut because some of the corroded pipe needed replacing. The refrigeration unit cooling the three MIC storage tanks was closed in the summer although Union Carbide specifically recommends the maintaining of MIC at temperatures below 5°C because it is an unstable liquid that quickly reacts to changes in temperature. A spare tank which should have been kept empty (in the event of an accident, the MIC from a faulty tank could have been transferred there), already had one ton of MIC sloshing around in it. The water curtain, a network of water meant to douse leaking gas,was functional on the night of the leak, but the water failed to reach the height at which the gas was escaping. And, finally, the vent gas scrubber, a tall, rocket-shaped tower which is designed to neutralize small leaks by flushing them with caustic soda, either could not cope with the large volume of escaping gas or probably didn't work at all. The sorry state of the safety facilities at UCIL was not altogether unexpected for the company had a relatively poor reputation for plant safety. However, its transgressions were viewed by local authorities with a benign eye, for conditions at other factories, including those run by the Government, were far worse.

In 1979, a major fire broke out at Carbide's alpha-napthol palnt, rendering it inoperable for weeks and causing heavy losses. No one was injured. However, the first major accident in the MIC facility occurred in December 1981, when a worker died after he was splashed with phosgene while trying to plug a leaking valve. It was this death that prompted Raj Kumar Keswani, the journalist, to begin his campaign against hazardous working conditions in the plant.

In February 1982, twenty-four workers were overcome by the effects of another phosgene leak. They were hospitalized. That

August, a chemical engineer was burned badly when liquid MIC splashed onto him. The last major accident before the December tragedy occurred was in October 1982, a few days after Keswani's first article had run. In that incident, a combined leak of MIC, hydrochloric acid and chloroform hurt three workers and set off a mild panic in Jayaprakash Nagar, the adjoining slum.

The October leak provoked the worker's union to plaster the walls outside the factory with posters in Hindi warning of the dangers to Bhopal. 'Lives of thousands of workers and citizens in danger because of poisonous gas', declared the posters. The posters weren't the sole warning of the dangers from the plant.

In May that year, a safety audit team from Carbide headquarters in the US travelled to Bhopal to check the safety status of the MIC plant. The team listed as many as ten major deficiencies in safety procedures the plant followed.[29] These included warnings of the potential for release of toxic materials in the phosgene/MIC unit and storage areas, either due to the failure of equipment like mechanical pump seals, operating problems, or maintenance problems (the use of substandard replacement materials was cautioned against). Other problems identified were the lack of water sprays in several areas and deficiencies in safety valves and instruments.[30] The high turnover in plant personnel was noted and commented upon. The team was also alarmed at practices such as maintenance workers being taken off jobs without being signed off and employees signing permits they could not even read.[31] It was noticed that although alarms and instruments were purportedly checked during shutdowns for correct setting, the plant did not maintain records of such checks.[32] Despite all these flaws the team declared it had been impressed with the plant's operating and maintenance procedures. 'No situations involving imminent danger or requiring immediate correction were noted during the inspection', the team wrote.

Its report was sent to Jagannath Mukund, who took over as the works manager in mid-1982. Mukund, a chemical engineer, who had worked for three years at the Institute plant, was to preside over the factory's slow death. During his tenure, he cut costs by reducing the number of workers at several units, including the MIC section. Workers and investigators say that the intensive training programmes for new recruits were diluted and staff from the alpha-napthol unit shifted to the MIC unit after a year's orientation. Most of the skilled engineers who had joined the facility at its start-up had left by 1984.

The report to Mukund, which is marked 'Business Confidential', was released to the press on 11 December 1984 by Union Carbide at Danbury.

Of particular signifance in the report of the three-man team is the stricture that workers should not be given responsibilities in toxic operations without sufficient understanding of safe operating procedures. The report indicated that not enough 'what if' training was conducted, which would have explained the entire system better to the workers and operators. This problem was never addressed and was still around two years after the report was first circulated. Most of the plant operators were high school graduates without degrees in chemical engineering. They spoke broken English and read it with difficulty.

The plant's operating manuals were only in English.

On 24 August 1984 R.K. Yadav, the general secretary of the main union at the UCIL factory, sent a four-paragraph letter to Mukund, the works manager, threatening to file a suit against the company before the state pollution board.

The company, Yadav alleged, was ignoring workers' complaints of rising atmospheric pollution and high noise levels in the factory. In laboured English, the letter stated that, 'you told us this is chemical plant and pollution is natural'. But, charged the union leader, who lived in a nearby colony with his wife and two daughters, pollution at the MIC unit and other insecticide producing sections of the factory, had reached an 'uncontrolled situation'.

Mukund's curt reply came the same day and ran into two pages. 'Our total plant complies with all requirements laid down by Central and State Governments with regard to water and air pollution. In fact, the State Pollution Board considers our factory a model for others to emulate'.

The works manager denied that pollution was rising anywhere in the plant and tersely told Yadav to 'stop making vague and general complaints which have no basis'. On the contrary, Mukund claimed, pollution levels had dropped.

Two years earlier, the union had complained to Tarachand Viyogi, then Minister of Labour in the state, of the lack of safety in the plant. Responding to opposition critics in the state legislature, Viyogi declared on 21 December 1982: 'It is not as if a great danger is posed to Bhopal by the factory. It is a Rs. 25 crore (about $20 million) investment, it is not a small stone that can be picked from one place and put at another'.

Three years after the disaster, there was still no agreement on what caused it.

In August 1986, Union Carbide claimed the tragedy was the result of an act of sabotage. It had spoken of sabotage as a possile cause of the tragedy as far back as March 1985 and also in its presentation before a New York court. But it had never specified the possible saboteur or how the act could have taken place.

The company made fresh details of its theory available in a long affidavit filed before the Bhopal District and Sessions Judge.[33] It also announced it was suing both the Central and Madhya Pradesh Governments for negligence leading to the tragedy. The 169-page statement was a rejoinder to the compensation suit filed by the Indian Government against Carbide in September 1986.

Stung by the accusations, the Indian Government brought a fresh affidavit before the court demanding that Union Carbide pay a minimum of a staggering $3.3 billion in compensation should the judge find it liable for the disaster.[34] It was the first time in nearly two years of litigation that the Government had publicly announced the amount of damages it was seeking. Its original suit had made no mention of the settlement it wanted and had merely repeated those allegations that it had first filed before a New York court in 1985.

Along with its demands, the Government denied any responsibility for the incident and rebuked Carbide for trying to pass the buck, saying that the sabotage theory could not exculpate the multinational.

CBI officials even claim that at least one manager tampered with evidence after the disaster and was seen draining water from pipes in the MIC area.

There are essentially two versions of how the accident occurred. One is the Government's supported by trade unionists, activists and the bulk of journalists; the other is Carbide's.

The Government's version of the causes that led to the tragedy relies heavily on a technical report[35] submitted by Srinivas Vardharajan, a senior government scientist who investigated the gas leak. Vardharajan published his report a full year after the tragedy. It differed sharply from Union Carbide's report which had been released nine months earlier. It differed too from one of Vardharajan's own earlier theories, first made public one month after the incident. He was quoted by the main Indian news agencies then as saying that as little as 'a pint' of water could have caused the reaction in Tank No. 610.

Eleven months later, he said that about 1,100 pounds of water would have been needed to set off the reaction. There was no doubt, however, that his investigation was more detailed than that of the company. This was unsurprising because the whole plant was opened up to his investigation, whereas the company's access was limited by the CBI.

According to the Vardharajan report, the residue in the tank 'clearly (showed) the entry of about 500 kilograms of water. In addition, metallic contaminants could have entered through a nitrogen line since nitrogen pressure had not been maintained for two months'. That pressure was stopped on 22 October, the report stated. Another way the water could have entered the tank was through the imperfectly monitored washing of lines near the MIC plant on 2 December. A metal disc (slip blind) which should have been inserted to prevent the water from entering the tank was missing, the report added.

In May 1985, alkaline water was flushed from pipes leading from the vent gas scrubber, the high structure meant for the neutralizing of MIC. The report said this strengthened the possibility that the water had backed up from the scrubber area into the two lines connected to it and down into Tank No. 610. The report picked on a special control valve as the defective part which possibly allowed the water to enter the tank.

The Vardharajan report said that MIC operators did not appreciate the problems that could result from water contamination. It also criticized the storing of large quantities of MIC as 'quite disproportionate to the capacity of further conversion of MIC....this permitted MIC to be stored for months together, without appreciation of potential hazards'. This factor, added to faulty design, poor choice of construction material and instruments, inadequate controls on storage systems and a failure to provide quick effective disposal, was cited as the main reason for the runaway reaction.

At no point, however, did the investigation speak of any failure on the part of the Government.

The reason for the disaster probably lies in a combination of various reports.

All are agreed on the point that water did enter the tank and react with the MIC stored in it.

The question is how?

This is a scenario advanced by government scientists, activists and journalists:

At 9.30 p.m. on 2 December the MIC plant superintendent, new to the job, took a worker to wash certain pipelines with water. The pipes led from the device that filtered MIC before it went to the storage tanks. The decisions to wash the lines was taken by the day-shift production superintendent because the Sevin section of the plant had restarted manufacture the previous week. The lines were also connected to the vent gas scrubber.

The lines were part of a vast web of complex pipes, valves and filters that would have stumped not just the ordinary layman, unfamiliar with the technicalities of the plant, but most of the plant's employees as well. In fact, most operators did not know much beyond their immediate responsibilities or understand how the whole factory and its systems worked. This was the danger the 1982 UCC safety audit had warned against.

The washing operation that night involved connecting a water line, closing an isolation valve located upstream, and opening several bleeder valves or drainage valves downstream. Usually, the lines were isolated by a slip blind, a metal disc inserted into a pipe to ensure that water did not leak through the isolation valve.

The disc was not in place and water began to collect in the pipes since the downstream lines, which should have been clear, were clogged. It rose past the leaking isolation valve into the relief valve vent header, one of two lines leading out of the MIC storage tanks. (In the event of a leak, this pipe was supposed to carry poisonous gas to the vent gas scrubber, where, theoretically, the stuff would be destroyed by a neutralizing burst of caustic soda. It was connected to a second line called the process vent header which carried excess gas released from the tanks, when they were pressurized with nitrogen, to the scrubber and the flare tower to be burned. After clearance from Union Carbide Corporation, UCIL had connected the process and relief headers, a major change in the original design.[36] The idea behind this was simple: should either of the vent header lines be out of commission, then excess gas could flow through the narrow U-shaped, connecting pipe, also known as the jumper, to the scrubber.)

From the relief valve vent header, the water flowed down into the tank, through the jumper line, the process vent header, which is usually open, and a series of valves which were either faulty or left open inadvertently. The water began to react with the MIC in the tank and pressure readings began to shoot up from the two pounds per square inch which had been recorded earlier in the day to ten pounds, then thirty. The temperature control was not working, and had not been for several months. This was an additional hazard as the refrigeration unit had been switched off in June. The temperature in the tank crossed 25°C, way above Union Carbide's recommended level of 5°C.

By 11.30 p.m., the workers had begun experiencing minor eye and throat irritation. V.N. Singh and two other workers went to check the source of the leak and located a leak of MIC and water from pipes near the vent gas scrubber, a good distance from the tanks. They directed a water spray onto the leak and returned for a tea break where the workers discussed the incident.

After the tea interval, Suman Dey noticed that the temperature gauge for Tank No. 610 showed a reading over 25°C, the top of gauge. He quickly checked the pressure level.

Forty and still rising. At forty pounds, Dey knew, the safety valve would pop. He rushed across to the office of Shakeel Qureshi, the supervisor on duty, told him of the soaring pressure and temperature and ran to investigate the situation at the storage tank area. The heat was so intense there that he could barely stay on top of the tanks for a second, before fleeing to safety. It was then that the cement above the tanks cracked; concrete shatters at temperatures over 400°C and the gas began escaping out of Tank No. 610.

A worker thought of pumping the remaining contents of the tank into Tank No. 619, which was supposed to be an empty backup, kept ready for emergencies such as this. But it was found to already contain some MIC and the idea was scrapped, for fear it would lead to a second eruption and an even bigger disaster.

Frantically, Dey tried to work the vent gas scrubber. The control gauges did not indicate a flow of caustic soda. The flare tower had been shut down for repairs, because the pipes leading to it were badly corroded. Even if it had been working, Dey and others say, the contact between fire and MIC could have caused an explosion. The gas was escaping at an initial rate of 180 pounds per minute and raced to a maximum pressure of 720 pounds per minute. Nothing that Union Carbide had devised for its Bhopal plant was capable of handling such a runaway reaction.

The water spraying system was turned on. This, too, was ineffective. The sprays reached about 100 feet off the ground, while the MIC was blasting out of the stack at a height of more than 120 feet. It enveloped the factory in a lethal mist 15 feet high and so thick that visibility was reduced to less than three or four feet. A worker activated the factory alarm system. With oxygen masks on their faces, the workers fled. Most of them made it out safely. Dey stayed in the plant, keeping upwind and occasionally donning his oxygen mask to visit the control room. He suffered no injuries, largely the result of keeping a cool head during the crisis. Shakeel Qureshi, the MIC supervisor that night, wasn't that lucky. He failed to locate an oxygen mask and panicked. He ran with Sushil Dubey, the Sevin plant operator, and tried to scale the fence behind the factory. Overcome by the fumes, the heavily-built Qureshi fell, breaking a leg. He was rushed to a hospital the next morning, kept in the intensive care unit and then placed under arrest.

The leaking valves and jumper line present one possible version of the events that led to the accident. Union Carbide has postulated another, more dramatic reconstruction of the events of 2-3 December 1984.

In May 1987, sources spearheading the company's investigation called me to present what they described as compelling new evidence to support Carbide's sabotage. Also present at the extensive briefing session was Steven Weisman, the *Times* bureau chief for South Asia.

This is Carbide's version of the incident:

Between 10.30 and 11.00 p.m. on 2 December, soon after the night shift took charge, a purported saboteur slipped into a narrow path that runs behind the tanks. He raised himself to the top of the tank, ignoring the steps nearby, after picking up a black hose lying in the lane. He fixed the hose to a water tap (known as a water drop in the plant). There were five such taps behind the tanks and they were used for flushing out solids and contaminants from the pipelines.

He unscrewed one of the two pressure indicator gauges, instruments used to check the air pressure inside the tank. The hose was fitted to the tank. The man turned the water on and walked away.

A short while later, an operator named M.C. Joshi, walking over from his post at the shut-down refrigeration unit, noticed the hose. The sequence of events after this is not clear in the Carbide scenario although it says that none of the three middle-level managers was at his post at the time. According to the company, Shakeel Qureshi and P.K. Jain, the shift supervisor on the fateful night, were smoking in the canteen. Smoking is banned in the MIC unit. K.V. Shetty, the shift superintendent, is logged as being at the main gate around that time to get some stores. The company believes that Qureshi and Jain were called back to their office—they shared a table there—by workers frightened by the scale of the reaction.

As the pressure and tension mounted in the control room area, one of the operators — the company believes it was Suman Dey, regarded as one of the most competent men in the unit — decided on a desperate measure to save the situation. He used pressure to move nearly one ton of the contaminated MIC out of Tank No. 610 to the chargepot, where in normal circumstances, it would have been mixed with other chemicals to make pesticide. The company says that Dey probably wanted to pump out the water, the chief contaminant which normally would have settled at the bottom of the tank. However, company officials postulate that Dey started pumping before too much water had entered the tank and so it was actually MIC that was pumped out. The internal reaction in the tank had increased air pressure and this higher pressure enabled the operator to force some of the stuff out from the bottom of the storage vessel.

The MIC that was pumped into the chargepot did not explode. It sat there although it had reacted to the water to a limited degree, attaining what Carbide called a 'greenish colour' and high levels of chloroform.

In an affidavit before the Bhopal district court, the company has named two operators, C.N. Sen and Joshi, as recording the transfer of MIC from Tank No. 611 to the chargepot between 10.15 p.m. and 10.30 p.m. on 2 December. But the company asserts that the actual transfer occurred between 12.15 a.m. and 12.30 a.m. on 3 December from Tank No. 610 and not 611.

In the later exposition to the *New York Times,* Carbide investigators spent most of their time trying to disprove the popularly held theory about water entering the tank from careless washing of the lines. The points made to debunk this theory were simple.

The sources said that a series of tests by hydraulic engineers and experts at two American laboratories proved that water could not have risen from the single closed pipe near the MIC structure where the washing was taking place. They say that log books and witnesses among the workers testify independently that all the safety valves were functioning and that precautions were taken for the water washing. The water was entering the few feet of piping marked for cleaning from the bleeder line and flowing out through three of the four outlets. The Carbide spokesmen, however, acknowledged that the fourth opening was malfunctioning and could not be opened. More important, they also acknowledged that the metal slip blind, which should have been inserted at the point of washing to isolate the rest of the pipelines, as factory procedures specify, was not inserted.

The force of the water at the inlet point was about forty pounds per square inch or about one-and-a-half times the velocity of water flowing through a garden hose. At that rate, the sources say, the water would have taken about forty minutes to an hour to fill the pipelines and then reach the tank. This is an important point because water, as even a beginner in physics knows, always finds its own level.

To rise above a particular point or to a particular high-level pipe, the water would have had to fill all the pipelines downstream before ascending. According to Carbide, more than 4,500 pounds or two tons of water would be required to fill all these pipes before even reaching the tank. It must be noted here that the pipes leading from the relief vent header and the process vent header to all three storage tanks flow down to the tanks. *But they are connected to the top of the relief and process headers.* This means that the water had to rise again (after filling more than 400 feet of pipeline, flowing through a large loop and then going up and down at least three places where the pipes rise or drop). 'To do that you would have to reinvent the laws of physics', says an engineer associated with the Carbide investigation.

Carbide even cites a CBI investigation to support its theory. When the CBI authorized plant officials to drill the process vent header, they found the pipe was bone dry.

Nor was any water drained from the subsidiary pipelines after the incident, the company says.

The company investigators say that the process header was itself slipblinded or closed at a point where it turned toward the MIC structure. This again meant that the water could not have gone beyond the slipblind to the scrubber.

They also question why water did not flow into Tanks 619 and 611 which are located before No. 610 and share the same safety system. According to them, the safety measures in these tanks were in place and untampered with. It was only 610 that was damaged, they allege, by a saboteur who was nursing a grudge over his transfer from the unit and who had been bypassed for a promotion.

Records and interviews with workers in the third shift that night indicate there was only one man who was 'transferred and bypassed'. He strenuously denies the charge although he admits to being bitter over his transfer. Other workers who were on the shift say the suspect was with them through the night.

Carbide has also alleged that there are conspiracies among the workers and the supervisors to prevent independent investigators from getting at the truth. The workers want to protect each other and the supervisors want to distance themselves as much from the event as possible.

But there are flaws in the conspiracy and sabotage theory. For one thing even Carbide admits that it will be difficult to prove sabotage except by scientific methods which rule out the water-washing theory.

For another there is no confession from the supposed saboteur.

Also to establish a breakthrough in its investigation and in its efforts to prove the sabotage concept, Carbide has relied excessively on one key informant. This is a man named Sunder Rajan, a plant engineer now relocated at Bombay, who testified before the Bhopal court that he had noticed the missing pressure indicator on the tank the morning after the incident.

Strangely, he made no mention of this stunning discovery when official investigators spoke with him soon after the event. Instead, that morning, he recalled, he simply went ahead and replaced it with another gauge.

'Why did Sunder Rajan take more than one year to remember that the p.i. was missing?' asked one Indian official in an interview. He pointed out that the disaster occurred in December 1984 and Rajan had not remembered the incident until February 1986, when he was probed by company investigators.

The original gauge has not been found. That is another flaw in the Carbide presentation, although one lawyer said that 'we knew we had hit pay dirt' when Rajan mentioned the missing indicator gauge.

Until that point, the company investigation had not turned up anything to support the sabotage theory. Sunder Rajan's remark started off a 17-month investigation which included more than 70 interviews with plant employees, mostly the second and third shift workers. The five-man team, which included four lawyers and a manager from the UCIL battery factory in Calcutta, pored over thousands of documents. They say they found strong suggestions that log sheets and timings had been tampered with.

Rajan himself has charged the CBI in court with intimidation and seeking to harass him into retracting his statement. The CBI denies the charges.

One problem facing anyone seeking to investigate the truth

about the matter is that all sides are depending heavily on oral evidence. This difficulty is reflected in varying accounts of a critical issue: the timing of the release of gases from the factory. The workers and supervisors say it took place about 12.15 a.m. to 12.30 a.m. on the night of 2-3 December 1984. The management says it occurred at 12.45 a.m. The question is critical because if one can settle on the timing when the gas vented from the tank, then one can deduce when water entered it and thus decide which theory is correct.

A question that needs to be asked at this point is whether it is more important to find out how the incident occurred or to secure justice for those who suffered.

Both are equally important for historic and immediate reasons. It would be as tragic as the disaster itself if the cause of the world's worst industrial mishap was not pinpointed.

But the question of who was liable for the disaster was sidestepped for a while when both sides announced they were negotiating an out-of-court settlement towards the end of 1987. The settlement plan raised an outcry and both sides backed off for a while, though it was becoming clear that unless a settlement was negotiated, the victims would not receive succour for many years to come. There is no question that those affected should be looked after as soon as possible. But the tragedy, if the case is never resolved in court, will be that the reasons that led to the night of destruction may never be conclusively proved. Also multinationals flouting the law in the future cannot be brought to heel with a precedent-setting judgement.

The Central and State Governments had dealt with Warren Anderson and his Indian aides. But their arrests did not solve a critical problem: what was to be done about the MIC that still remained in the other tanks? Should it be neutralized through the scrubber by pushing little bursts of it through the caustic soda solution? Should it be repacked into smaller containers and shipped back to Danbury? Or should it be coverted into pesticide by using the existing facilities?

The decision was to be taken by Vardharajan, the chief of the Council for Scientific and Industrial Research (the Government's main pool of industry-related scientists) who was heading the team of scientists investigating the disaster. Vardharajan, an elegant dresser, who favours dark, two-piece suits, held doctorates in organic chemistry and bio-chemistry from the universities of Cambridge and Delhi. He also had behind him a successful career as a manager of several government corporations.

Vardharajan visited the plant, familiarized himself with the equipment and the processes, checked the storage area where the disaster occurred, and spent hours questioning Mukund and other factory officials. He also consulted the American technicians who had come in.

He asked for the release of Mukund and the four other technical managers, saying that their presence was crucial to the safe disposal of the gas. Without them, he said in a letter to the state's Chief Secretary, 'the process of neutralizing the MIC could be extremely dangerous and lead to a second tragedy'. Mukund and the others with him were set free on bail. So were Keshub Mahindra, the UCIL chairman, and Vijay Gokhale, its managing director.

Vardharajan's letter represented a rare case of a government seeking the release of men it had accused of serious criminal offences. By taking this approach, he was admitting that his team would be helpless without UCIL's expertise.

Vardharajan also supervised experiments at the plant involving MIC. These were conducted with company staff, to get an idea of the various possibilities that could have caused the reaction in Tank No. 610 as also ways of safely disposing of the MIC left in the tanks. For these tests, the team used small amounts of MIC stored in stainless steel containers.

After consulting the Americans and Mukund, Vardharajan decided that converting the remaining MIC into pesticide was the safest option. But, he told the Delhi crisis cell that, the plant needed stocks of phosgene, the gas which acts as a stabilizer for MIC and is a key component in its manufacture. Water selectively reacts first with phosgene, which functions as an inhibitor. Phosgene prevents polymerization, a chemical reaction which strings together existing molecules into a chain of longer molecules to destablize the liquid. Vardharajan felt the phosgene would be needed to stabilize Tank No. 611, which then contained at least 15 tons of MIC.

The crisis management team in Delhi quickly identified a phosgene producer, Atul Products, located in Gujarat. Army helicopters secretly flew in two lots of phosgene from the gas producer at Valsad. Ultimately, the gas was never used.

The secrecy surrounding operations at the plant heightened the existing tension in the area. Thousands began streaming out of the city a second time, by bullock cart, tonga, bus, train, rickshaws and pushcarts, on foot and in cars. Vehicle-owners waited patiently to fill up at petrol stations. This time, the migrants locked their homes and shops, organized bundles of clothes and food before leaving and thronged the railways and bus stations to get out. The fresh exodus forced the Government's hand. On the twelfth of December, the Chief Minister declared that the plant would convert the remaining liquid MIC into harmless pesticide.

'All safety measures will be adopted', Arjun Singh said in a statement. He identified the most important of these as the spraying of water from a helicopter. This was derided as a cosmetic measure by critics. The start-up process was to begin on 16 December and it would be named 'Operation Faith'. Nine camps were to be located in the city's schools and colleges for those who wanted to move out of their homes. Special buses would ply from specific colonies to transport residents to these sites. Camp inmates would be fed two meals a day and would receive medical treatment. But they needed to bring their own plates and cups, mattresses, pillows and sheets.

The Government then announced the closure of all schools and colleges until the end of the detoxification, confirming fears among the populace that even it was unsure of the process's safety. Special buses transported people, their household goods and even animals. Fleeing residents sat on top of buses because there was no space inside. In the trains, divisions between first class and second class passengers were forgotten. Labourers and their families, with their little bundles of food and clothes, sat next to prosperous women in shimmering silk saris and gold jewellery.

The railway authorities did not intervene.

On 16 December, a clear winter day, helicopters hummed overhead and crop-duster planes swung in the air letting loose a fine spray of water over the plant and its neighbourhood. This process was supposed to be a safeguard against leaks: once in the atmosphere, MIC reacts with water to become non-toxic dimethyl urea.

Arjun Singh arrived nearly an hour after the process of MIC conversion had begun. He had been electioneering the day before at a distant district. Inside the Union Carbide plant, Vardharajan and his crew were not taking any chances. With hard hats and walkie-talkies, they communicated with Union Carbide personnel and with each other. But they did not really interfere with the working of the factory.

Prior to the operation, they had gone through the entire plant closely, looking for flaws. The rupture disc was replaced on the safety valve above the MIC tank and a fresh filtration system was introduced on the nitrogen line. Operators continuously monitored pressure levels. The nitrogen supplied by a neighbouring air products firm was checked for purity by a scientist.

A scientist sat in UCIL's control room while Vardharajan himself shuttled between the administrative block and the plant, a distance of nearly half a kilometre. Wet tarpaulins covered the tower near the scrubber to neutralize any leak. Engineers put up sacking on bamboo poles on the walls of the factory facing Jayaprakash Nagar. These were doused with water by tanker trucks and the city fire brigade. The tankers also poured water on the streets and pavements. Mobile medical teams were in position, ready to move at short notice.

Yet, all these precautions meant little to the inhabitants of the old city. Even the Chief Minister's declaration that he would stay at the plant during the neutralization carried no weight. 'He has a plane, a helicopter, he can fly away at any time, we can't', said one resident of Jayaprakash Nagar as he packed his bags and bundled his family into a waiting bus. About 10,000 people, many of them too poor to pay their fares to remote hometowns and villages, moved to the temporary camps. In all, more than 300,000 people, most of them inhabitants of old Bhopal, fled, shuttering the area and turning it into a deserted town.

The tension inside the plant mounted as the start-up began. Would nitrogen injected into the second tank push the MIC out of the storage tank into the chargepot, the one-ton container where it would mix with other chemicals to form Sevin? Or would it fail, as efforts to pressurize and clear Tank No. 610 had failed before the disaster?

The liquid flowed without faltering to the chargepot. As the temperature and pressure remained stable, the technicians heaved sighs of relief. The conversion process could continue. Alpha naptha and carbon tetrachloride flowed into the reactor vessel, transforming an estimated three tons of MIC into Sevin.

Scores of curious but apprehensive citizens with handkerchiefs covering their faces, a few politicians, and a gaggle of Indian and foreign reporters, television crews and photographers looked on and listened. In the evening, on the lawns outside a police official's bungalow, Vardharajan told waiting newsmen that the first day had been a success.

The process continued a full week, instead of five days as predicted. The delay was caused by the discovery of additional amounts of MIC lying in the tank being cleaned and the adjoining one. This was much more than the 15 tons that the Indian scientists were told existed in the plant's vitals.

Obviously, UCIL's inventory of its stocks was as poor as its leaky system. In all, a total of nearly 24 tons of MIC was processed into pesticide, including 1.2 tons in stainless steel cylinders.

As the operation continued safely, the people of Bhopal began streaming back into the city. Markets began to open, homes were unlocked, government employees and workers in private firms began reporting for duty. The relief camps were shut and their inmates bussed home. The Government announced plans to rehabilitate gas victims, pay some cash relief, arrange free rations and medicines.

Slowly the city began to revive.

# The Fight To Save Lives

When the gas rolled into Bhopal railway station, Nissar Ahmed, an electrician with the railways, was sleeping in a storeroom off the main platform. Awakened by screams outside the room, Ahmed went out to check. Scores of porters and passengers were lying on the platform, dead or dying, gasping for breath as the thick mist enveloped the station. About 600 were evacuated from the area when help finally came several hours later.

When his eyes began to burn and his throat to hurt, Ahmed rushed to a bathroom and splashed his face with water. He repeated this several times. It was a move that saved his life.

Two days later, he travelled to his wife's home in Jhansi, 250 kilometres to the north, vowing never to set foot in Bhopal again. To his concern, and that of his family, Ahmed developed signs of mental instability. He became mody and depressed. He lost his temper with the children and beat them often. Speaking to his wife of his mental confusion and disorientation, he said: 'I never know what to do'.

In January 1985, the electrician travelled to Bombay's King Edward Medical College where S.R. Kamat, a chest specialist, had begun treating victims of the gas disaster. Kamat was regarded as India's foremost authority on industrial diseases.

In Bombay, Ahmed's condition was diagnosed as a case of neurotic depression. He had also suffered extensive lung damage.

Years after the accident, doctors in Bhopal were still reporting fresh cases of neurotic depression. *'Dil main ghabraat hain, neen nehi aati hain, sir dukh ta hain'* (I feel deeply worried, I don't sleep and my head aches), Hamida Bee, a bulky burkha-clad woman who lives in Jayaprakash Nagar, told a psychiatrist in July 1986.

Indeed, mental health problems have been a major fallout of the disaster.

However, psychiatric problems were far from the minds of the doctors on 3 December 1984. They were more concerned with stopping hundreds of blinded, gasping people from dying on them.

Nothing in his experience had prepared Harihar Trivedi, a medical specialist, for what he faced at the Hamidia hospital the night of the disaster. Trivedi, lanky and *paan* chewing, was among the first of the senior doctors to reach the hospital after getting word of the tragedy.

The place was in utter chaos. An estimated 10,000 victims were already in the hospital complex, a sprawling, evil-smelling network of corridors, wards and offices, sitting and lying on the beds, tables, chairs, anywhere they could rest or lie down. Every moment, trucks and cars and ambulances would screech to a halt outside the emergency ward entrance, unload more patients and rush away. A large number of the patients were critically ill.

Trivedi and others including N.R. Bhandari, the portly superintendent, did not try to organize the situation. Already, most of the hospital's staff, and students of Gandhi Medical College, the state's biggest medical institution, were fully stretched. As elsewhere in the city, where rescue and relief operations were primarily non-formal, at the Hamidia too, the response was voluntary, instinctive, swift and quite effective. The doctors were facing a problem they had not been trained to tackle and so what they actually did in the crisis was very commendable.

Until then, neither they nor medical history knew anything of the effects of MIC. One American expert said later that all that was known of the chemical and its effects could have filled a couple of pages.

The doctors at the Hamidia, naturally, knew of no antidote and settled on symptomatic treatment.

On duty that night in the emergency room was Mustaq Mohammad Sheikh. The first patient with burning eyes arrived at the hospital at 1.15 a.m., forty-five minutes after Tank No. 610 leaked. In another five minutes, Dr. Sheikh was swamped by hundreds more who milled around in the small emergency ward. Help was soon on the way. Scores of residents and interns, worried by personal breathing problems brought on by the gas, rushed to the hospital to see what had happened. They were pressed into service along with nurses, junior doctors and staff members. They washed their eyes and faces with cold water, as Nissar Ahmed had done at the railway station, and used atropine eye drops for relief. They also placed wet cloths and handkerchiefs over their faces for further protection.

One batch of young doctors, injured and frightened by the gas, fled their hostels. But most of the others stayed on.

At other hospitals in the city, there were similiar scenes. In the Kasturba Hospital, run by the Central Government's Bharat Heavy Electricals Limited, health officials treated thousands outside the building.

At the Hamidia, quick-thinking interns realized that the existing facilities were about to collapse under the pressure.

So a number of them moved out of the building and, armed with injections, eye drops and water, set up outdoor clinics with benches serving as counters. Those complaining of burning in the eyes—most people were coming to the hospitals with their eyes shut in pain—were told to flush their eyes with clean water.

They were then given eye drops to reduce photophobia (which causes blurred and distorted vision), before other ailments were examined. Patients were instructed to use eye drops two or three times a day in the initial week. If this had not been done, experts say, a membrane would have grown between the undilated tissue around the pupil and the cornea, causing blindness.

Yet, despite all the doctors did, victims complained that they still could not see properly. A large number of patients also suffered opacity or clouding of the cornea, which manifested itself as white spots. After a couple of years though the major eye problems had diminished, some of the effects of the gas continued to linger. A report in September 1986 said there was fresh concern among doctors at the Indian Council of Medical Research over possible blindness setting in.

The scientist who reported this development, C.R. Krishnamurti, is a toxicologist and the head of a government committee which is studying the toxic effects of MIC on humans, animals and vegetation. He felt the blurring of vision among eye patients could lead to eventual blindness. However, Dr. Krishnamurti also said he could not furnish details on how many people would actually go blind as a result of renewed corneal opacity.

His report was immediately challenged by other specialists at the ICMR who sent down a team to investigate patients with eye problems. They found no alarming jump in corneal opacity cases and reported a total of eleven per cent among 2,000 patients. ICMR's deputy director-general, Usha Luthra, said that as there never had been a specialized survey immediately after the disaster with sophisticated equipment it was difficult to compare problems of that time with current ailments, which are being studied with spanking new imported equipment.

Whatever the present opinion of the medical community, the eye treatment was initially regarded as among the more successful of the various medical efforts. People used water instinctively, without being told of its effectiveness. Tapas Ganguly, a superintendent at the telegraph department, splashed his eyes with water when he felt the stinging in the air. It was what the patriarch of an entire family in Jayaprakash Nagar also did. According to a health worker who visited the family the next morning, the old man had dipped several towels in water and ordered his people to place it over their faces. All of them survived, while entire families in their lane were wiped out.

Water was what L.D. Loya, the doctor at the Union Carbide plant, prescribed when the first frantic calls came through from Dr. Sheikh at the Hamidia. The gas, Sheikh was told, was non-poisonous. The only way to deal with it was to place a wet cloth over the face.

Such superficial treatment was useless for the seriously injured. For Loya knew, as did other key Carbide personnel who had read the company manual on MIC, that the chemical was deadly. The Bhopal manual made it quite clear that the material was both toxic and flammable and overexposure could lead to fatal pulmonary oedema, or filling of the lungs with water. Basically, this meant that people would drown in their own body fluids, caused by the reaction of MIC to moisture in the lungs and respiratory tract. Also, the dissolution of lung and respiratory tract tissues would destroy lung capacity. In some cases, as much as half or more of lung capacity was destroyed. This cut down the ability to breathe. The result was a painful death.

To counter breathing problems, the doctors instinctively used a combination of oxygen, bronchodilators, diuretics and strong doses of corticosteroids. The aim was two-fold: reduce immediate inflammation and prevent secondary infection.

The high doses of steroids led to several complications, among them diabetes and obesity as the drugs upset the body's hormonal balance.

One problem on the first day of the disaster was that there were not enough oxygen cylinders to go around. Appeals for help went out to private clinics through the city. UCIL flew in cylinders and masks from Calcutta and Delhi and stocks of cortisone drops. The Indian Government and other centres in Madhya Pradesh air-freighted medicine and personnel. By the fifth of December, an additional 400 doctors had arrived in the city to supplement the health effort.

On that day, word came from Awasia, the multinational's medical director in Institute, West Virginia, on possible antidotes to the effect of MIC. Listening to a radio broadcast of the disaster as he drove to work early the previous day, Awasia instructed his office to send the following telegram to the doctors at Bhopal: 'If cyanide poisoning is suspected, use Amyl Nitrite. If no effect Sod. Nitrite 0.3 gms and Sod. Thiosulphate 12.5 gms'.

The brief message was to mark the beginning of an ugly controversy.

As the death toll began to mount, Heeresh Chandra, at the Hamidia, began conducting autopsies. He found definite evidence, he said, of hydrogen cyanide in the bodies he examined. The blood was cherry-coloured, the lungs were red and so were the brains. A man with a military moustache and chiselled features, Chandra declared to an unbelieving medical community in Bhopal that the gas victims were dying of cyanide poisoning and recommended that patients be injected with sodium thiosulphate, the recognized antidote.

He himself had begun using the drug and so had others at the morgue, involved in the autopsies. As they cut into the corpses, the gases released often left the doctors drained and breathless. So the forensic team, too, had turned to sodium thiosulphate. It gave them, Chandra remarked afterward, much relief and they were able to work more efficiently.

Meanwhile Awasia, Carbide's health director, had reached Bhopal. He met the Hamidia's senior doctors, including Chandra and N.P. Mishra, the dean of the medical faculty. They questioned him on his message about thiosulphate. 'Will this gas liberate hydrogen cyanide', Chandra asked. 'No', replied Awasia. Chandra then pointed out that Carbide's own manual said that at certain temperatures MIC does release hydrogen cyanide. 'Yes', replied Awasia. 'At what temperature does this happen?' asked the Hamidia specialist. At 437°C he was told.

Chandra knew by then that the concrete shield above the MIC tanks had cracked. Concrete cracks at temperatures above 400°C, although some experts believe the shield cracked because the steel tank below bulged upward during the internal chemical reaction. Chandra asked Awasia what components in MIC would release cyanide. Awasia was evasive, saying he was still collecting data.

On 8 December, Max Daunderer, a respected West German toxicologist, travelled to Bhopal at the invitation of Vulluri Ramalingaswami, the tall, ascetic-looking head of the Indian Council for Medical Research, the country's top medical research institute. Daunderer conducted clinical tests on the victims, listened to Heeresh Chandra and came to the same conclusion: the patients should be administered sodium thiosulphate, to counter cyanide poisoning. He tried out the drug on a number of cases. While one patient recovered spectacularly, another died triggering ugly charges of medical malpractice by local doctors. Daunderer flew out of Bhopal soon after, concerned that his personal safety was at risk.

But he warned of the following three-stage complications: eye, skin and lung problems would continue in the first week; then, between three or four weeks after the incident, victims would report disruption of the central nervous system; the final and long-term problem would involve more complications of the nervous system with chances of paralysis and other problems.

In fact, before Daunderer arrived in Bhopal, the Hamidia's doctors were reporting their first cases of mental stress and nervous disorders. Nine cases, sharing a common complaint—convulsions—were recorded by Bhandari, the portly hospital superintendent. Thus, although the predominant impression after the disaster was of a city where everybody coughed, the malaise was far more serious. Unfortunately for the victims, the early historic medical effort was soon forgotten and replaced by bureaucratism, politics and personality clashes.

First, the state's medical authorities in Bhopal overruled Heeresh Chandra and went ahead with symptomatic treatment. Union Carbide could be faulted here. First, for not giving adequate information about the toxicity of the chemical and making available any record of treatment for MIC poisoning and, second, for asserting that cyanide and other toxic chemicals (such as phosgene) had not escaped in the leak.

'It is not phosgene, it is not phosgene, it is not phosgene', snapped Awasia at a 14 December press conference, where he was grilled by angry local journalists, some of whom were injured by the leak.

Two other American specialists, sent by Union Carbide to assess the devastation, then intervened and adroitly steered Awasia and Loya, the local company doctor, out of trouble.

The specialists, Hans Weill, a professor of pulmonary medicine at Tulane Medical School, and Peter Halberg, an opthalmologist and profesor at New York Medical College, praised the Hamidia effort. Halberg said the quick treatment of the eye problems meant that there would be no blindness in Bhopal as a result of the leak. Weill was more cautious. 'No one can say with any assurance what the future holds', he said. Noting that this was the first mass exposure to methyl isocyanate, Weill added that a break in oxygen supply could damage the brain, the nervous system, the heart and even foetuses of pregnant mothers. Also, as many patients had left the Hamidia against medical advice, follow-up treatment would be patchy. 'The problems will continue over the long term, and become chronic', Weill predicted.

Medical history had no real parallels with the Bhopal tragedy.

Weill had studied cases of American workers exposed to chlorine leaks in Florida. But medical researchers had shied away from working and experimenting with MIC because of the compound's high toxicity. Most of the tests had been on rats, mice rabbits.

In 1964, four humans were exposed by West German toxicologists to the gas for periods of one to five minutes. At least one of the research group was a toxicologist himself, a German named Georg

Kimmerle.[1] At a low level of MIC exposure (0.4 parts per million), Kimmerle reported, the subjects had no problems. When the dosage was increased five times, their eyes began to water and irritation spread to their throats and noses, At 21 ppm, the exposure became unbearable. There the experiment stopped. It is not known what the long term effects of the exposure were on the four persons but the experiments prompted American industrial scientists to recommend a total exposure limit of 0.02 ppm over a eight-hour work shift. This is one-fifth of the exposure limit recommended for phosgene. The Bhopal victims were exposed, says Dr. Chandra, to levels 'hundreds of times above these limits'.

MIC, is part of a family of isocyanates, the three best known of which are used in the polyurethane industry to produce foams, insulation and coatings. Their names are tongue twisters: toluene diisocyanate (TDI), diphenylmethane-4.4-diisocyanate (MDI) and polymethylenepolyphenyl polyisocyanate.[2] The effects of TDI and MDI on humans have been studied quite extensively.

The major health problems experienced by people exposed to both these compounds were breathing difficulties and sensitization. Isocyanates are known to attack the respiratory system, eyes and skin. Thus, they can wound the lungs and bronchial tracts and cause long term eye damage. The exposure standard for TDI in the United States is now 0.005 ppm.

Unlike the other compounds, MIC is not widely used and its function is basically that of an intermediate in the production of pesticides. Thus, despite the best efforts of Bhopal's doctors, they faced tremendous limitations in terms of their knowledge of MIC's effects on the human body. This led to what became known as the sodium thiosulphate controversy. Chandra, who is the head of the city's medico-legal centre, pressed his case for sodium thiosulphate's use as an antidote despite opposition from N.P. Mishra, the dean of medicine, who said that the results of tests did not establish an absolute truth. K.W. Jaeger, a toxicologist from the World Health Organization, refuted Chandra's position as the medical establishment held out against the cyanide poisoning theory.

'Under no circumstances shall sodium thiosulphate be given unless it is correctly and conclusively proved that there is cyanide poisoning', declared M.N. Nagu, the director of health services of

Madhya Pradesh. By then, a motley band of activists, ranging from a fiercely anti-establishment scientist to a film actress, doctors and writers, had become involved in the controversy. They insisted that sodium thiosulphate was the right treatment and that the Government was being pressured by Union Carbide not to go ahead with it.

An ICMR report in mid-December 1984 urged an early resolution of the controversy. Yet, a month later, the Government had still not decided which way to rule on the cyanide issue. Finally, after the close observation and monitoring of patients who had received the drug, the ICMR met with Mishra and several other doctors from the Hamidia in February 1985 and declared that sodium thiosulphate injections had proved useful. The council said that four types of patients were to receive the injections. They were identified as people who suffered from breathing, gastric and neuromuscular problems; others who complained of recurring ailments; those who had gone into a coma or were suffering acute lung damage after the incident; and those with a history of death from the MIC leak in their families.

The policy was reviewed in April when the ICMR said that the injections had shown good results in a number of cases and should be continued. Strangely enough, despite findings by researchers that this treatment was effective up to a point, the Madhya Pradesh Government did not use it on a mass scale until another month had passed. It was only after activists in Bhopal and Delhi stepped up their demonstrations, rallies and protests that the Government acted. In retrospect, its reluctance to more on the issue is inexplicable.

In May 1986,[3] when the ICMR published an update on conditions in Bhopal, it noted that the effectiveness of sodium thiosulphate injections was dropping. This was probably due to the depletion of the poison in the body's systems. Yet, in that report, the ICMR raised another controversy. It said that eighteen months after the disaster, the gas that leaked from the Union Carbide plant was still unknown. 'The nature of the gas that descended on the unfortunate citizens of Bhopal on that fateful night remains to be established, but the suffering of the afflicted continues unabated,' the report said.[4]

One thing was clear, though—the MIC had degraded into an

array of gases and compounds during and after the reaction in Tank No. 610.

Although the sodium thiosulphate treatment is continuing, medical research has now uncovered a vast range of ailments in victims of the tragedy which are unlikely to be cured. The damage to the lungs is the most widespread and is regarded as permanent. The lungs can be repaired to some extent, doctors agree, but they cannot be restored to full health.

Those who must do physical work for a living cannot carry heavy loads any more. 'Which passenger wants a coolie who can carry only half the weight and walks like an old man?' asks Sukh Ram, who was injured by the gas. Their physical disability means that incomes for men like Ram have dropped and they must struggle even harder to earn a living. In addition, they have to waste a lot of money at the myriad clinics that have come up in Bhopal to dispense medicines for every possible ailment.

There are antacids for stomach problems, anti-depressants for the depressed, steroids for those with internal inflammation, vitamins for general strength and, of course, sodium thiosulphate injections.

Pregnant women were among the worst-affected by the gas. In one of its studies, the ICMR investigated a total of 2,566 women who were pregnant at the time of the disaster. The survey revealed the following findings: in the women surveyed, there were 355 spontaneous abortions or three times the national average; 56 still births or a four-fold jump over the national average; 91 infant mortalities within a week of birth or twice the national average. The infant mortality rate among babies who completed one year of age on March 1986, and who were born to mothers exposed to the gas, was 110/1,000 live births. The national figure for infant mortality in urban areas is 65.2/1,000 live births.

Warnings of foetus damage were sounded in a series of separate reports issued by activist doctors and other specialists who monitored smaller groups of women. These groups did some of the best work on various health issues. They were also unanimous in their opinion that the Government was incompetent.

Rani Bang, a young gynaecologist who visited Bhopal, with her husband, a few weeks after the accident, studied a total of 218 women, including 114 women affected by the gas.[5] Her findings, even at that early stage, were ominous. As many as sixty-six gas-afflicted women reported a white vaginal discharge; twenty-seven spoke of excessive menstrual bleeding, while thirty-five complained of pain in the lower abdomen.[6] Two even declared that their husbands had become impotent.[7] While these findings could not be regarded at the time as authoritative, Dr Bang reported that of the twenty-seven lactating women in the affected group, sixteen or nearly sixty per cent, spoke of a drying up of breast milk.[8] She later acknowledged that this could have been caused by the physical stress and mental trauma touched off by the tragedy.

In March 1985, Mahesh Yeshwantrao Rawal, the head of the department of obstetrics and gynaecology at Bombay's K.E.M. Hospital, conducted a survey of 198 women living within eight kilometres of the Carbide plant.[9] More than half this number reported gynaecological problems.[10] Several of these difficulties were common to the Bang report—heavy bleeding by nine women during menstruation and excessive vaginal discharge. Pregnant women also complained to Rawal that the foetuses were moving less than earlier. Rawal thought, however, this was a 'subjective feeling'.[11] He said it could have been induced, as in the case of Bang's patients, by a fear among pregnant women that their children could be abnormal. This was a justified concern, as Rawal discovered when he returned to Bhopal later and conducted a second series of tests on a different group of women. 'We feel that in 58-60 cases, there will be some deformity', Rawal said after the study.[12] He, of course, did not mean just gross abnormalities such as limb deformities, but was referring to underweight babies and disturbances of the central nervous system as well. Studies are continuing on the effects of the gases on children born after the leak. There is still no authoritative data and secrecy shrouds the project. Despite all this, after the initial chaos, lack of direction, and sharp personality clashes, the entire medical effort appears somewhat better organized now. There are computerized lists of persons being treated under the ICMR's twenty-five research projects, there is new equipment imported by the Government and donated by people wanting to help. This was not the case in the summer of 1985, when confusion, inefficiency and poor book-keeping marked the health effort. Doctors were casually prescribing pain-killers, antibiotics, steroids, antacids. There was no documentation or proper records kept of patients, many of whom went from one clinic to another in search of more pills. Different groups of doctors advocated different methods of treatment.

But although there is a fair degree of improvement in the manner in which the medical effort is being conducted, there are still a number of problems. For instance, researchers faced a unique obstacle. Many of their subjects were non-hospital-based and poor. These people were becoming increasingly reluctant to spend a whole day every week or so undergoing tests and anwering questions. A day in the hospital meant the loss of a day's earnings, something they could ill afford. So they often would just stop appearing at the clinics on the appointed days.

According to doctors and health officials, research and treatment are still separate endeavours in Bhopal. The medical effort is suffering because soon after the leak, patients were treated in several centres, and were often overdosed. Also, as records were seldom kept, follow-up treatment was not possible. As a result, many patients have apparently lost faith in the public hospitals and are turning to private clinics that a majority of them cannot afford.

A new 100-bed hospital was opened in November 1986 for patients with chronic ailments, near the Union Carbide plant, but it can hardly meet a fraction of the actual need. A relief official has estimated the actual number of patients suffering long-term effects at 25,000. Many of them are children.

One hundred days after the leak, the K.E.M. survey declared that large numbers of children exposed to the gas were unable to play games because of chest pain and breathlessness.[13]

According to Ashwini Siyal, a paediatrician at the Hamidia, mothers of children born after the disaster often complained that their offspring showed little interest in food and tired easily.[14] The women have less breast milk and wean early. Added to this is the problem so endemic to developing countries—malnutrition brought on by poverty. An enormous but unsystematic relief programme has not helped.

Ishwar Dass, the relief commissioner, says that nutritious bread is distributed every day, six days a week, at special centres in the city for children under six and lactating mothers. Both groups also get free health care. But set against this are things like the drop in income facing most affected families and, more important, their subsistence level existence. Awasia, the Union Carbide doctor, had warned about this when he had visited Bhopal soon after the leak. The extent of recovery, Awasia said, could depend on a victim's level of nutrition. 'Someone who is malnourished may very well be more damaged', was his contention.

As things stand, the number of those seriously afflicted is still enormous. Three years after the disaster, the mental health problems that Max Daunderer, the German toxicologist who left Bhopal in a hurry, had predicted were still surfacing.

In the summer of 1986, a group of field staff and researchers began a systematic survey of badly and moderately affected areas to identify psychiatric cases, The team noted that as many as twelve to fifteen per cent of adults in the families surveyed were suffering from mental stress and depression.

Indeed, the continuing problems of mental health are the greatest cause of worry. It is, as Ramalingaswami, the ICMR chief, puts it 'a virtually unknown area'. A lot of criticial time has already been lost. 'There may still be many who are suffering and who are not coming to us'. says Dr. Santosh Tandon, a psychiatrist who is treating patients with mental problems.[15] Initially, the treatment of mentally ill patients suffered a setback as they were being treated for what were regarded as physical problems: breathlessness, fatigue and headache. Until the middle of 1985, the Hamidia did not have a separate psychiatric section and doctors were just not trained to spot such problems. Mental stress patients were administered symptomatic drugs. The doctors could not be blamed. They did not know better in the absence of comprehensive medical information.

In order to remedy this, Tandon and another psychiatrist were brought in by the ICMR from Lucknow's King George Hospital, which has a large psychiatric department. They began giving on-site training to health workers in order to enable them to identify psychiatric cases. That was when mental health treatment first began to be organized.

By then the leak was more than eight months old. 'The long

term problem will continue to be in the area of psychiatry, perhaps more than any other area', says Kamat, the chest specialist at Bombay. 'There is no way of knowing how this will be resolved'.

Whether the MIC could have a carcinogenic effect on those exposed to it—by acting on the DNA—is being debated, although researchers said initially this appeared unlikely. Specialists at the Hamidia are also looking at possible long-term genetic damage by monitoring children born after the accident. But American immunologists who have collaborated with Kamat at the K.E.M. Hospital say that blood samples of Bhopal victims examined at the University of Pittsburg have shown the development of antibodies to MIC.[16] This means that the immune system has recovered from the initial attack on the proteins and enzymes and developed a plan, as it were, of counterattack to rid the body of the unwelcome invader.

But all the news on the medical front is not grim. Quite apart from the upgrading of the general medical effort, the work done by individuals such as Kamat of Bombay's K.E.M. hospital is highly praiseworthy. Unlike the initial relief effort at Bhopal, Kamat's has been an organized one from the start. Thus, when the medical community in Bhopal was divided by controversy and mistrust, Kamat, his deputy A.A. Mahashur, and others, put together a systematized series of tests on more than 100 patients. Most of these were railway employees who were given special passes to travel free of charge for three-monthly check-ups. They were paid a special incentive fee of Rs. 200 by the K.E.M. for every visit.

'They get a free trip, a free check-up, money to spend and a visit to Bombay', says Kamat, fifty-four, who is a ninteen-year veteran at the hospital.[17] His clinic is a three-storey building set behind the Gothic facade of the municipal hospital. In a room that is chilled by air conditioners, the doctor explains why his method has produced results, at least in reducing lung damage. Indeed Kamat's treatment appears to have had better success than that of, say, Dr. Mishra (a Bhopal doctor who has also worked a great deal with lung-damaged victims); of the 1,085 patients Mishra saw regularly, the lung ailments had 'stabilized' whereas among Kamat's patients as many as 90 per cent had actually recovered to varying degrees. More important, their health continues to improve. 'We have shown that MIC's lung ailment is not a slow killer but is slowly reversible', Kamat says. How has he succeeded, where so many others have failed? The answer is that whereas other doctors were haphazard in their treatment and not particular about follow-up, Kamat had dealt with cases like Nissar Ahmed, the electrician, and other railway workers, from December 1984 on. They were seen regularly. Without embroiling himself in the sodium thiosulphate controversy, Kamat set about his own diagnosis and treatment.

This involved simple physiotherapy—easy exercises to strengthen lung capacity and develop physical energy, the occasional use of anti-depressants and the regular use of two drugs: aminophilyn and levimisal. Aminophilyn, a popular drug among asthmatics, reduced lung inflammation and levimisal strengthened the body's immune systems. In the long run, says Kamat, it was levimisal which played the key role in recovery.

Also pivotal to his research and diagnosis was a close scrutiny of the only previous large scale exposure to an isocyanate. In 1967, a fire at an industrial plant in England released large amounts of TDI, toluene diisocyanate. Thirty-five firemen were exposed to it. Two studies of that incident reported that during a four-year follow-up period the subjects developed symptoms which, Kamat noted, bore an uncanny resemblance to the Bhopal victims he was examining.

In the initial hours after exposure to TDI, the firemen had eye, nose and throat irritation, nausea, vomiting, abdominal pain and diarrhoea. Chest symptoms were at their worst on the first three days and subsided later.[18] 'Simple lung function measurements showed a persistent decline', said a study.[19] Acute and chronic breathing problems were noted in a small number, and also lack of resistance to secondary infections like asthma and pneumonia.[20]

'It took the firemen three years to improve, we saw major improvements in our patients in one-and-a-half years', Kamat says. He is not given to modesty, perhaps understandably so. 'We are better organized than the Bhopal people, we have better credibility'.

The State Government could have done with some of that credibility in the first few months after the disaster. Barring the individual initiatives of people like Puri, the police chief, and others at the actual time of the tragedy, the Government's bureaucratic response to the disaster was delayed, unimaginative and predictable. For a start it was unable to piece together enough information on the death toll. Over the first three days the fatalities listed essentially reflected a count of those bodies which had been brought to the mortuary at the Hamidia.

On the first day, when reporters placed the number of fatalities at about 450, the Government reported a figure of about 350. The next day, the relative counts rose to 1,200 and nearly 600. That day, many government inspectors were making their first visits to graveyards and the cremation ghats to look at the registers of those buried and burned. While the media's figure steadied first at 2,100, and later 2,500, the Government's assessment of the death toll did not rise beyond 1,750 until much later. In July 1986, the relief commissioner declared that his department had confirmed 2,100 deaths. (Later, the Government alleged 2,850 deaths in its compensation suit). Of this figure, the dependants of 400 victims were still to receive their relief payments of Rs. 10,000. Only eleven of those who died were insured against death and disability, a reflection on the impoverished state of most of the victims. The insurance agent who dealt with the cases said he ran through them in less than two hours and dispensed a total of Rs. 400,000 to the victims' next of kin.

An additional 36,000 persons, with a monthly income of less than Rs. 500, who suffered physical disability because of exposure to the gas, were paid Rs. 1,500 each as compensation by the Government. However, 3,000 others, whose cases were confirmed, did not turn up to accept payment till late 1986. The situation is still patchy.

Dass, the chief relief official, says that complaints about lack of payments first began surfacing in 1985 (the decision to pay compensation to the disabled was taken that April). A fresh survey produced an additional 32,000 names of people who said they hadn't got their compensation payments yet. 'Most of these people have names that are common to the lists we already have, but they are spelled slightly differently', says Dass.[21] Thus, 'the computer finds it difficult to match the lists'. The untangling of the process has taken much longer than anticipated. Towards the end of 1987, as the two parties began talking of a settlement, of the nearly 5,30,000 claims filed, only 80,000 had been processed and it was estimated that a further four years would be needed to clear all the claims.

Indeed, the whole distribution of free food and other relief measures had been criticized on grounds of corruption and worse. Even Dass, the relief commissioner, accepts some of the charges. 'Forgeries of counter-foils (used to attest to the fact that the bearer has been affected by the gas) have taken place; so have cases of impersonation and people have got money two times over', he says.

In a well-intentioned and highly-publicized move to feed those who were too ravaged by the effects of the gas to work, the Government announced in the second week of December 1984, free rations to those with ration cards. It then issued another 21,000 ration cards to those who didn't have them. A short while before the December 1984 general elections, Arjun Singh's administration declared that it was extending the free food scheme to the entire *jhuggi-jhopri* (hutment dwellers) population in Bhopal. At best, it was an ill-concealed political manoeuvre.

A staggering 500,000 people were given a free ration package of 12 kilograms of rice and wheat, half-a-kilogram of cooking oil, and half-a-kilogram of sugar per adult. Each child was given 200 millilitres of milk every day. Between the start of the program and the time it ended in January 1986, the Government spent a total of Rs. 450 million on relief and rehabilitation. Of this figure, Rs. 200 million went on free rations. The money was a loan from the Central Government to Madhya Pradesh.

Finally, the burden became too great for such generosity. The Government stopped the free rations programme abruptly, setting off protests by voluntary groups and activists. 'We felt that enough was enough and people could not be supported indefinitely on handouts', Dass says. An unfortunate fallout of the whole effort was that many of the vicitims, particularly those who were physically weakened and mentally depressed, lost the will to work during their year on the dole.

Also, the rampant corruption among officials in charge of the relief effort eroded its efficacy. Says Ram Singh of Jayaprakash Nagar, 'The food inspectors are millionaires now.' Activists and local residents say that the ration store owners too made a lot of money, holding back rations and selling it at high prices on the black market. Such stories are widely accepted as true. It is a pity the Government's relief effort was not as effective as it should have been but then corruption and mismanagement are not the best aids to charity.

The activists were angry, as activists usually are. Soon after the disaster, hundreds of volunteers descended on Bhopal to monitor the Government's efforts, organize health clinics and assess the damage. From the beginning, they went after the Government with a vengeance.

By their lights, everything the Government did was either wrong, shortsighted, incompetent or intended to benefit Union Carbide. The Government, the groups declared, did not care for the gas victims but only about saving face because it was as guilty as Carbide in that it had allowed the plant to come up and then had not held to high safety standards.

Their words had sting, much truth and much appeal for the afflicted. Thus, the Government, already struggling with the disaster, its reputation and medical controversies, was faced with a newpublic embarrassment.It was not that the activist groups posed a political threat to the administration. But they did draw public attention to the Government's scars at an incovenient time.

From December 1984 onward, the radicals battled the State Government. They took Heeresh Chandra's side on the sodium thiosulphate controversy, and demanded mass use of the drug. This issue, popularized through rallies and demonstrations, was one of the most significant contributions of the activist movement in Bhopal. The only problem was that it was so grudgingly accepted by the ICMR and the State Government, that it slowed the drug's implementation and, by extension, its efficacy. To be fair to the ICMR, it did declare within weeks of the disaster that the controversy over cyanide therapy needed to be settled urgently. However, factional disputes in the Bhopal medical community delayed the resolution of the issue.

The sodium thiosulphate controversy was only the first of several issues the activists took up. A number of the groups, including the *Zahreeli Gas Kand Morcha* (Poison Gas Incident Front) and the *Nagrik Rahat Aur Punarwas Samiti* (Citizen's Relief and Rehabilitation Committee) concentrated on the medical effort and urged that it be better organized. Prominent among these activists were people like Anil Sadgopal of *Kishore Bharati,* an activist group based at Hoshangabad in Madhya Pradesh, and his wife Mira, a doctor. Another star was Suhasini Mulay, an elegant film actress from Bombay, whose jeans and classy, manicured looks contrasted vividly with the impoverished people she sought to organize. She later helped produce a documentary film on the disaster and its aftermath which ran into trouble with the film censors.

Others, such as a Bombay-based group known as the Union Research Group, set up a trade union relief fund which looked after small rehabilitation schemes for gas victims. Lawyers also banded together to prepare cases and briefs and advice for those wanting to go to court. The Participatory Research Society in India (PRIA), based at New Delhi, produced pamphlets in Hindi and English as did the Delhi Science Forum, another well-known association of activists. PRIA's Rajesh Tandon says his organization is working with others, especially the Centre for Science and Environment, perhaps the country's best-known and respected environmental forum, in setting up what he calls a 'national coalition' of trade union leaders, rural organizations, doctors and others in hammering out a common front to deal with the issues that continue to arise out of the tragedy in Bhopal. Such a co-ordinated effort is, however, yet to emerge and there are still different groups concentrating on different issues.

One other major issue on which the activists came together after the sodium thiosuphate controversy, was the disorganized and ineffective relief effort. Claims were not being honoured, their processing was taking too long, were the major plaints, all too familiar to those with experience of the country's governmental machinery. Rallies and demonstrations were organized on this issue until the administration finally cracked down on the activists, arresting their leaders when they threatened to block train traffic. The voluntary clinics were closed, their files damaged, their documents seized. Some activists even complained of beatings by the police.

The groups were shaken and demoralized. Many pulled out, concerned about their physical safety.

Some came back briefly on the anniversaries of the gas leak, addressed meetings and took part in rallies. They left soon after. Today the activist presence is far more low-key, almost token, say officials and local residents. Bombay-based trade union organizers still visit the city and continue rehabilitation programmes, especially for women.

That the Government was still ultra-sensitive to activists and their criticism of its projects was seen in the way it dealt with two cases in Bhopal in September 1986. In the first instance, the police arrested an activist who had taped the proceedings of a meeting of the ICMR. They would have arrested a second man had he not managed to escape. The arrested activist, Gautam Banerjee of Calcutta, was charged under the Official Secrets Act, a law that has been used against spying and other treasonable activities. Banerjee hardly falls into such a category, say other activists and reporters, who see harassment of activist groups as the real motive behind the move. Officials in New Delhi were embarrassed by the Bhopal police's zealousness to protect the reputation of the State Government, especially as the incident occurred a few days before the Indian Government filed suit against Carbide.

The second arrest which stirred protests was that of David Bergman, a twenty-year-old Englishman, who had cycled all the way from England to Bhopal to raise funds for the gas victims. In the city, he set up a crèche for children in the gas-hit neighbourhoods. He was detained for overstaying his visa and also accused of spying, first for Union Carbide, and later for the Central Intelligence Agency. The *Indian Express* described the charges as indicative of 'police paranoia'. The case against him was dropped and he returned to Britain. But he was relentless in his criticism of the administration's handling of the relief projects. The Government, he said had 'done nothing for the gas victims'.

But the activists' approach too was flawed. First, they adopted a very high profile approach, concentrating on media coverage and mass rallies. This had three effects: one, it helped them score an impressive win on the sodium thiosulphate front. That remains their most tangible achievement. Two, it led the gas victims to believe that public protests were the only way to achieve results.

Three, the public response, the media exposure and the favourable resolution of the sodium thiosulphate issue made the groups overconfident which in turn led to their overestimating their strength. What the activists did not realize was that by confronting and alienating the Government on every possible front, they had made an implacable foe. When it saw its interests being directly affected and the activists' role grow into an embarrassing political challenge, the state apparatus acted in the only way it could—a police crackdown to crush dissidence.

There are lessons to be learned here for public service organizations.

One is that there are times for a high-profile approach. And there are times when a low-key movement pays better dividends. The initial drive and publicity over sodium thiosulphate served a purpose. Thereafter, the groups should have concentrated not just on mobilizing the people but on starting imaginative schemes which would have made a real difference to the lives of those in the gas-ravaged areas.

Another factor that reduced the activists' effectiveness was that many of the activists were outsiders, non-Bhopal residents. Although the locals appeared unable to achieve anything much, the involvement of a man like Mahesh Buch, a housing expert, for instance, would have paid dividends. Buch is a man obsessed with the issue of urban housing. He says that the most tragic part about the disaster is that the gas-affected areas still look the same or even worse than at the time of the disaster.

According to Buch, the Government should have simply razed the old hutments in Jayaprakash Nagar and built a new modern housing colony with proper sewers, water and power connections and schools and playgrounds for the children. The residents of the colony could have been given the material to build the houses themselves and even been paid for their labour in constructing the new colony. Instead of pressing for and arranging for these things, which could have made visible difference to the lives of the victims of Bhopal, the activists got bogged down in mobilizing people in a somewhat pointless fashion.

It isn't only the activists who have been critical of the Government's relief effort. In January 1986, in a long note to Motilal Vora, the man who succeeded Arjun Singh as Chief

Minister, a senior Goverment official wrote, 'During my field visits...a notable feature has emerged regarding the condition of a large number of families, which continue to live precariously. I have been told about such families who may have got the relief as per the scales prescribed but it has not helped them in a very meaningful way'.[22]

The official, who was associated with relief projects, said in his note that the key issue in the rehabilitation process was the question of mental health. Urging a review of the projects, he said, 'mental health considerations must constitute the essential ingredient in such a reassessment'.[23]

Later on his in report he became more scathing. 'I do not think that the present level of health infrastructure available can ever expect to cope with an enormous problem like this'.[24] The Chief Minister never responded to the report.

Health officials in Delhi echo the Bhopal official's sentiments when they say they are frustrated by the slow progress of the medical effort at Bhopal. 'We can't run things from here by remote control', says one top official at the ICMR. 'The problem is that the local administration doesn't seem really interested in getting things done', she adds.

Another ICMR official says that if the Bhopal medical community had adopted Dr. Kamat's methods of working—organizing patients, maintaining decent records, monitoring patients closely, ensuring efficient follow-up—if not the treatment he used, the health situation could have been transformed. 'The trouble was that they let inter-departmental feuds and controversies supplant scientific investigation for too long', the official says.

There is no doubt that if controversies such as the sodium thiosulphate one had not been allowed to get out of hand, a lot more victims would have been better today. In every way the Bhopal victims were and continue to be affected by forces quite beyond their control.

# The Legal Battle

As 1987 drew to an end there was news of India and Union Carbide negotiating an out-of-court settlement. The man who nudged the two parties to the bargaining table was Judge Mahadeo Waman Deo, the silver-haired district judge of Bhopal. A man fond of quoting Charles Dickens in his rulings, the judge held that the resolution of the case outside the courtroom was in the best interests of the gas victims.

Initially, both sides stuck to their original positions, but with less acrimony than had characterized earlier meetings in the United States. Union Carbide offered $500 million as compensation. (This was broken up into $2000 a year for 10 years for the dependents of the 2600 dead—UCC's figure; the Government's is 2850—$1000 a year for 10 years for the chronically ill, and $500 as a lump sum for the slightly injured. Compensation to the Govenment, local corporations and resettlement fees would be around $160 million). The Indian Government said no to this offer and demanded a compensation figure closer to $615 million. But before the settlement could be clinched, and alleviate the pain of the victims, a furious outcry broke out in the press and other forums. Former Chief Justice P.N. Bhagwati (who was responsible for a unique concept of managerial liability in the case of polluting industries, as part of a ruling against Shriram Fertilisers in December 1986) along with 39 other eminent persons, in a letter to the Prime Minister, demanded that any settlement must 'insist upon an admission of liability by Union Carbide' saying that such a decision could be a bench-mark for other countries in the Third World. The signatories claimed that an out-of-court settlement (would) involve a dereliction by the Government of (its) duty.'

But the signatories who are also pressing for multinational liability (as mooted by Justice Bhagwati when he was Chief Justice) should take note of the fact that such a novel theory will find few takers in an American court to which Carbide could rush. This could delay the case as much as 20 years, according to a Carbide source.

What the victims will gain out of a long-drawn-out battle is questionable. As S. Sahay, former Delhi resident editor of the *Statesman,* writing in the *Telegraph* of 17 November 1987, said, 'what is involved in the Bhopal case is both principle and money...if both can be had, well and good, but if a choice has to be made it should be in favour of money...' A cabinet minister told me that the Government wanted to settle quickly and that it was prepared to face opposition over a possible deal with Carbide. He and Judge Deo stressed independently that the victims had not gained through the court battle; only the lawyers had profited.

But before the case landed in the Bhopal District Court presided over by Judge Deo, it travelled thousands of miles, first from India to the United States and then back.

Judge Deo was the fourth to hear the case. One of the previous judges had been injured in an accident and the third, G.S. Patel, was found to be a claimant himself in the case. This embarrassing discovery in early 1987 followed a series of decisions that Judge Patel had given, forcing Union Carbide to pledge a minimum of $3 billion in assets to meet any eventual compensation claims. Carbide lawyers considered applying for a mistrial on the basis of a conflict of interest since Patel had a stake in the case, but decided against it.

The journey to the Indian courts began on 12 May 1986, when Judge John F. Keenan of the Federal District Court of the South District of New York announced that he was sending billions of dollars of damage claims against Union Carbide Corporation back to the Indian courts. These, he said, were the most appropriate forum to decide the various issues involved. Judge Keenan imposed three conditions on the multinational saying that the decision to send the case back depended on Carbide's acceptance of the points.

These were that the corporation and not just its subsidiary would be governed by Indian jurisdiction, that UCC would accept a fair Indian court verdict on the issue of compensation and that the company would abide by the same pre-trial disclosure proceedings that it had accepted in an American court.

Basically, Judge Keenan had accepted Union Carbide's plea of *forum non-conveniens*—that New York wasn't the appropriate place for the legal battle because most of the documents, litigants, evidence and witnesses were in India and it would be a waste of time, money and energy transporting people and material to the United States when the case could be heard more easilly in India.

The victims in Bhopal were dismayed and depressed, their hopes of big American money promised by US lawyers, who had visited Bhopal after the disaster, and signed up thousands of people wanting compensation, vanishing; the Indian Government, the chief litigant and representative of the injured and dead, was surprised but philosophical, taking solace in the judge's conditions on Carbide; the American lawyers, representing victims privately, were upset and later filed an appeal against the ruling as it meant they would be shut out of any eventual settlement. They had earlier been upset by India's decision (ratified by an Act of Parliament) to make itself the sole legal representative of the people of Bhopal. Naturally, the Indian Government could not impose such a concept on an American court. But when the tide began to turn towards trying the case in India, the freelance American lawyers became seriously worried. With reason, as subsequent events showed, for they lost an appeal against the Keenan ruling in the American courts. 'It's the end of the road', one of the American lawyers said.

Union Carbide itself, while welcoming the decision to return the case to India, was taciturn about the court's conditions, saying it needed to study them.

No one appeared happy with Keenan's ruling although it marked the virtual end of more than a year of courtroom battles, secret negotiations between the company and the Indian Government, selective leaks to the news media (with both sides trading bitter charges) and a judge's painstaking but failed efforts to resolve the tangle before it came to trial.

Within days of the announcement, experts in New Delhi were examining ways to bring a giant to judgement in India.

The legal battle with Carbide began scant hours after the leak, when Prime Minister Rajiv Gandhi visited Bhopal and pledged to seek compensation. Vasant Sathe, then Minister for Chemicals, also declared that he expected the company to respond with compensation at American levels. But for weeks the Indian Law Ministry was uncertain about how it was going to prosecute a multinational, which was head quartered thousands of miles away.

Inspiration came from an unlikely source: the dozens of American lawyers, derided as 'ambulance chasers' in their own media, who flocked to Bhopal to see how far they could take the American law of liability. They successfully sold dreams of millions of dollars (to be extracted from Union Carbide) to thousands of victims if only the case was tried in the United States.

'At first the air was filled with poison. Then with lawyers', intoned the *Washington Post* in a sharp editorial soon after the disaster.[1] The newspaper criticized the American lawyers for starting an unseemly race for mega-million-dollar compensation suits. The *Post's* well-intentioned advice, which seemed partly to stem from a sense of outrage that Americans were insensitively exploiting a tragedy, had no effect on its targets. An essay in the *New York Times* by Robert Stein, a lawyer involved in environment issues, called the rush of lawyers both 'insulting and inappropriate'.[2] That didn't make much of a dent either.

To the lawyers, the situation was a virtual gold mine, in terms of money and reputations to be made. They were not particularly bothered about whether they were viewed as rapacious and insensitive. They stood to gain as much as one-third of any settlement arrived at in the United States. Besides, they felt, only an American court could hand out damages that were really substantial—in the lawyers' estimation, the settlement could eventually run into several hundred million dollars.

The lawyers aside, the main reason the case hadn't immediately been filed in India was because of a lack of precedent. No industrial accident of the magnitude of Bhopal had ever occurred in the country. And the issue was not just industrial safety but also damage to the environment.

Although India had both a Central Pollution Control Board (to check toxic emissions in the air and water) as well as state-level boards, the number of prosecutions and convictions these bodies had managed to clinch was pitifully small. This was because the law was weak, with light and easily bailable sentences.

In fact, environmental activists preferred to take erring industries to court on the basis of public interest litigation, a system of seeking justice on issues that were crucial to public well-being. But this concept was new to Indian jurisprudence and had been barely tried in the field of environment. There were only a dozen or so lawyers who specialized in it (the architect of public interest litigation was former Chief Justice P.N. Bhagwati of the Supreme Court.) Besides the concept was only practised in the higher courts with their greater powers and reach, rather than in the district and other smaller courts.

Another difficulty in filing the case in India was presented by the fact that the company involved was a subsidiary of a worldwide industrial giant. The question was what sort of legal action would be really effective? Indian courts (the Indian legal experts felt, and here they were in concordance with the American lawyers) would have no control over the coffers of multinational, located in the United States when the issue of compensation arose. The local courts could only ensure that the Indian company paid.

Yet another factor for looking to the United States was the precedent of large compensations for victims of industrial negligence and accidents. American courts had handled mass tort actions and handed down strong rulings. (Tort is the law governing redress sought through a civil action for injury, wrongful act or damage.)

Two major cases decided in the US in the 1980s drew a lot of legal attention. One involved the payment of $200 million to former American servicemen in Vietnam who had been injured by exposure to Agent Orange, a devastating defoliant. The money was paid out by the manufacturer because the judge ruled the company owed product liability for causing the injuries. The second was in 1982, where a judge in California ruled that A.H. Robins, the manufacturer of an intrauterine device known as the Dalkon Shield, which had proved ineffective in numerous cases, should pay $38 million in compensation to scores of litigants. He said that prosecutors had proved the liability of the company in manufacturing an unsafe device—many women complained of physical problems after using the device—and the Dalkon Shield was taken off the market.

Product liability was the key point of both the rulings: the courts felt that a firm manufacturing a product that caused personal injury and damage was legally responsible for its defective product. The Dalkon Shield case and the Agent Orange case were regarded as benchmarks in determining product liability in the United States. Interestingly, the Dalkon Shield case was won by the firm eventually picked to represent India in the US—Robins, Zelle, Larson and Kaplan.

There was another reason for the Indian side (and American freelance lawyers) seeking a trial in the US. In that country, the doctrine of strict liability meant that lawyers had only to prove their clients had been hurt by defective products. Under Indian law, one had to prove negligence, a far more complex task.

While this debate went on, American lawyers continued to fly into Bhopal with printed compensation forms for the victims, most of whom were illiterate and could barely sign their own names. A number of the victims simply inked their thumbs and pressed a stain onto the forms, confirming that they agreed with a document they did not really understand. One of the lawyers was seen handing out currency notes to people outside his hotel. For this act of munificence, his clients signed compensation claim forms in Hindi and English nominating the lawyer their representative. They agreed to pay fifty per cent of their share of any settlement to him. Of course, if there was no settlement they weren't bound to pay him a cent. This arrangement is known as the contingency fee in the United States and is standard legal practice in that country.

The law of torts was to be the vehicle of action. A leader in tort cases, Melvin Belli, flew into India after filing a $15 billion suit on behalf of two gas victims in Charleston, West Virginia. One of them was Rehman Patel, the railway official on duty on the night of the disaster. In Delhi, Belli met with Indian legal officials and private attorneys before travelling to Bhopal. In Bhopal, he was given a great deal of importance with bureaucrats, politicians and reporters eagerly soliciting his view and advice.

But Belli wasn't the first on the scene. That distinction went to John Coale, of the Washington firm of John Coale Associates. Coale landed in Bhopal barely a week after the disaster. His partner, Arthur Lowy, an elderly man who wore a trilby and limped, and he, lost no time in hiring a aggressive young man named Raman Khanna, who was associated with the local Congress Party. There were two others in the Coale team: an Indian tailor from the Washington area named D.S. Sastry, and Ted Dickenson, a lean, tough-looking private investigator, whose brief was to check the credibility of the claims put to the team.

Sastry spoke Hindi and helped set up press interviews and meetings with local officials. But Coale also grew to depend on Khanna, a lightly-bearded man who was known as a fixer. The American lawyer had some experience with liability cases. He had represented twelve of the fifty-two American hostages in the 1980 crisis at the US Embassy in Teheran.[3] In another case, he had sued Brazil for injuries suffered by a young man who was shot by the son of the Brazilian Ambassador in Washington. He won that case against the Brazilian Government.[4]

Coale and his associates initially declared themselves pleased with the progress they had made in Bhopal. They had signed up thousands of claimants and had a letter from the Mayor of Bhopal, R.K. Bisaria, appointing the firm the city's legal representative.

But competition ensured that Coale's gains were short-lived. A rival team of lawyers from Santa Monica, California, produced Deep Chand Yadav, chairman of an influential committee in the Bhopal municipal corporation and Bisaria's political foe, who declared that there was no accord on a legal nominee for the city. The issue remained unresolved.

Sastry later fell out with Coale. In the spring of 1986, he filed a suit against the lawyer saying Coale had offered Bisaria a hefty bribe and an expenses-paid trip to the United States.[5] Coale angrily denied the charges but acknowledged that he had hired Bisaria as a consultant after the Mayor had relinquished office in 1985.[6] He said he'd paid Bisaria $8,000 in fees and expenses. But that was not all.[7]

According to Sastry's affidavit, Bisaria had once complained to him that he was owed $5,000 for helping get death cases signed up for the Washington law firm.[8] He said he had been promised $200 for every such case.[9] Sastry claimed he had paid the former Mayor $1,800 in cash and a further Rs. 50,000 later.[10]

More sleazy details involving other lawyers slowly emerged, pointing to a network of fraud in the whole matter of claims. Raman Khanna says that several hundred people that the American lawyers signed up just didn't exist.[11] Because they came in a hurry and wanted to get their work done quickly, Khanna and other local intermediaries hired bands of college students to sign up clients. Soon Khanna began noticing the same names appearing in his lists. When he questioned the students, they admitted to putting false names to the claims.

'They were getting Rs. 25-to-30 per case they brought in', he said. 'They would simply write out any name that they could think of and sign the claims themselves'. The city's taxi drivers got to know the groups of students well. Whenever a new lawyer arrived in town, they would direct him to the students and collect small commissions from both sides.

Coale said initially that he represented around 50,000 claimants. When the signing spree ended a total of twenty-nine firms had signed up between them 148,000 victims demanding up to $100 billion in damages in Judge Keenan's New York court. Elsewhere, more claims were filed in state courts in Texas, California and West Virginia.

The Indian Government, although it disliked the noise and brashness of the American lawyers, thought that they had the key to an effective platform to sue Union Carbide. Three officials who were crucial to developing India's strategy were K. Parasaran, the Attorney-General; B.S. Sekhon, the country's Law Secretary; and Shyamal Ghosh, a Joint Secretary in the then Ministry of Chemicals, which is now part of the Industry Ministry.

The Government decided to make the Chemicals Ministry the nodal point for the handling of all Carbide-related matters for it was the arm of government which had dealt most frequently with the company. Ghosh was later to become chief negotiator and trouble-shooter for the Government.

In January 1985, Parasaran flew to the United States to consult with a number of firms and lawyers and the Indian Embassy over the modalities of filing against the corporation in New York. His discussions were inconclusive. By then, the Government was worried by the tens of thousands of compensation cases that had already been filed in the United States and in Bhopal. There was even one in the Indian Supreme Court which sought to implicate the State and Central Governments for alleged negligence leading to the accident. Pressure was mounting on the Government because some of the American cases were to come up for hearing in April. If it was not to be shut out of the picture, New Delhi had to move quickly.

First, the Government issued a Presidential ordinance declaring itself the sole representative of the gas victims, effectively cutting out private lawyers. The ordinance was converted into an Act of Parliament when the new Lok Sabha, the ruling Lower House of the People, convened for its first session (in February 1985) after the death of Mrs. Indira Gandhi. The Government based its Act on the concept of *parens patriae,* the unique idea that a nation-state had the right to exclusive legal representation for its people. It also waived court fees for seekers of compensation, eliminating what would have been an excessive burden on the victims.

While the primary aim of this move was to eliminate the American lawyers, the Government conceded that Indian law would not hold good in an American court and wrote in a clause allowing private representation in a foreign court at the expense of the compensation-seeker.

Getting the ordinance into the statute books was not a problem because Parliament was overwhelmingly dominated by Gandhi's supporters who held 401 of 542 seats.

Meanwhile, Attorney-General Parasaran in a second swing through the United States in March (along with Sekhon, the Law Secretary) had finalized the lawyers who would present India's case. The winners were Robins, Zelle, Larson and Kaplan who had fought and won the celebrated product liability case involving the Dalkon Shield. Robins, Zelle was a low-profile firm from Minneapolis, on the Great Lakes, located in America's Midwest. The firm did not have the glitz and glamour of legal firms from New York or the West Coast, but it did have a reputation as a tough, successful disaster and product liability specialist.

Indeed, Robins, Zelle was a pioneer in personal injury litigation in the United States with nearly fifty years of specialization in the field.[12] It had successfully sued the giant film studio, Metro Goldwyn Mayer, for a fire in the MGM-owned Grant Hotel at Las Vegas, Nevada.[13] It had won cases against polyurethane manufacturers and numerous other negligence suits.[14]

While the selection process was on, a federal judicial panel, comprising federal judges from different parts of the country, met in Washington to consider the scores of cases filed in the United States by various law firms against Carbide. On 6 February 1985, the Judicial Panel on Multidistrict Litigation ruled that eighteen lawsuits pending in different courts be transferred and consolidated in a single court—that of Judge Keenan in Manhattan. New York, the panel said, was the logical place for the hearing of the case because Carbide was a New York corporation and its head offices at Danbury were closer to New York than to Chicago or Los Angeles. Consolidation of the cases had become necessary, the panel declared, in order to eliminate duplicate discovery, prevent inconsistent pre-trial rulings and conserve the resources of the parties, their counsel and the judiciary.

The consolidation of an eventual sixty-four lawsuits in New York did not answer another question. Who would represent the victims and argue their cases before the judge? There were almost as many lawyers as cases. They were bitterly divided and intensely combative.

Judge Keenan announced his solution: a three-member executive committee comprising two attorneys who represented private clients and one the Indian Government. The white-haired judge gave the lawyers a week to sort out their differences. If they couldn't settle the leadership issue in that time, he said, he would name the two lawyers himself. April 23, the deadline, passed without a resolution of differences. Keenan declared his choice—Stanley Chesley of Cincinnati and F. Lee Bailey, a well-known trial lawyer of Manhattan, would represent private interests. Michael Ciresi of Robins, Zelle would be India's man. While India pursued its legal strategy, it also readied itself for a negotiated settlement with Carbide. There was nothing contradictory in this. The Indian Government felt that pressure by way of the courts would only strengthen its position in negotiation.

At the end of March 1985, Rolf Towe, a vice president and treasurer of Union Carbide Corporation, arrived in New Delhi. He had been authorized by Warren Anderson, the company chief, and others to come to India and search for a solution before the legal battle became either too protracted or bitter. Towe's meetings marked the beginning of several rounds of secret talks that continued, at long intervals, for a year between Parasaran, the Indian Attorney-General, Anderson and others.

According to Vijay Gokhale, the Indian subsidiary's chairman, the Government expected far too much from Towe, who was stunned by the Government's demand: it wanted his company to hand out a settlement of more than $one billion. The sum was several times higher than anything Union Carbide was prepared to pay. The company offered to pay $100 million (to be paid out over thirty years) which, with interest, would grow to $200 million. The Indian Government dismissed the offer. They were also annoyed by the company's proposal to conduct a virtual audit of the way the money would be spent. 'The kind of offers made showed a lack of perception of the dimenstion of the disaster', B.B. Singh, the Secretary in the Chemicals Ministry, later told a small group of reporters from American newspapers and agencies. He emphasized the fact that the Bhopal tragedy was not an accident but a disaster. The parent firm was also trying to distance itself as much as possible from UCIL, he said.

The gap between the two protagonists appeared unbridgeable.

On 8 April, Robins, Zelle filed an affidavit against Union Carbide in Judge Keenan's court accusing the company of controlling UCIL 'from cradle to grave' and alleging Carbide responsibility for the mishap. Towe returned to New York.

On 26 June 1985, American lawyers representing private parties filed a joint suit against Union Carbide, using language similar to that in the Indian plaint.

For the moment, the time for talking was over.

The suits against the Corporation were unique in several aspects. For one thing, they made no mention of damages sought saying 'surveys and numerous scientific and medical studies are currently being conducted....and at such time that the surveys and studies are completed, plaintiff will allege a figure for compensatory damages sustained by persons with claims'.

The Government was to repeat these words in the complaint filed against the company at Bhopal more than one year later. Although much had happened at Bhopal in the intervening period, with relief efforts being slightly better organized, the lack of imagination in the phrasing of the new plaint told much of the real story. Granted, the size of the task was unprecedented. Yet the same lethargy that appeared to invest the Indian rehabilitation effort now dogged the Indian Government's efforts to put its case together against Carbide.

The precise number of deaths—2,344—was not filed until October 1986 (this was later raised to 2,850 ). The number of seriously afflicted was placed at 30,000 to 40,000. A specific figure of compensation was not mentioned too until then—the Government wanted a minimum of $3.1 billion. (Even this sum was not properly arrived at and Judge Deo, the last judge to try the case in Bhopal, embarrassed the Indian Government by saying he wanted a detailed break-up of the figure).

But these difficulties notwithstanding, India propagated a new legal theory at New York, called Multinational Enterprise Liability. In its statement before Judge Keenan, the Government said, 'Key management personnel of multinationals exercise a closely-held power which is neither restricted by national boundaries nor effectively controlled by international law. The complex corporate structure of the multinational, with networks of subsidiaries and divisions, makes it exceedingly difficult or even impossible to pinpoint responsibility for the damage caused by the enterprise to discrete corporate units or individuals.[15]

'In reality, there is but one entity, the monolithic multinational which is responsible for the design, development and dissemination of information and technology worldwide...[16]

'Persons harmed by the acts of a multinational corporation are not in a position to isolate which unit of the enterprise caused the harm, yet it is evident that the multinational enterprise that caused the harm is liable for such harm.'[17]

The Government added, 'A multinational corporation has a primary, absolute and non-delegable duty to the persons and country in which it has in any manner...undertaken any ultrahazardous or inherently dangerous activity.' The Government felt Carbide had breached this duty by 'taking unacceptable risks at its plant at Bhopal, and the resultant escape of lethal MIC from that plant...Union Carbide failed to provide that its Bhopal plant met the highest standards of safety and failed to inform the Union of India and its people of the dangers therein. [18] This failure meant the corporation was primarily and absolutely liable for any and all damages caused by the gas.'[19]

In its fifty-point plaint, the Government also faulted the corporation on nine specific issues relating to plant conditions. Among other things, the Government alleged that the corporation had recommended, encouraged and permitted the storage of MIC in dangerously large quantities; that it had not provided for an intermediate facility between the production plant and storage tank, creating the potential for a contaminant to enter the tank; that the tanks had not been isolated properly or well chilled, and that they lacked adequate temperature alarms; that the vent gas scrubber had been defectively designed and poorly maintained. The Government charged the company with failing to share elementary medical information during the crisis. It also maintained that UCC had not given its local subsidiary information about a review conducted at the Institute MIC plant in September 1984 which had spoken of the chances of a runaway reaction in the MIC manufacturing process under certain conditions.

The Government also accused Union Carbide of falsely asserting that the Bhopal plant was designed with the best information and skill. Rather piously, the Indian side claimed that it had relied totally on the multinational for assurances of safety (a poor reflection on its insecticides department whose facilities are inadequate to test the long-term effects of chemicals and assess the need for toxic chemicals in the manufacturing process). The Government described Union Carbide's assurances of providing the best technology as 'unlawful, wilful, malicious and reprehensible' and (that its entire operations) were in 'deliberate, conscious and wanton disregard of the rights and safety of the people of India'. In its summation, the Indian Government said that compensation should adequately meet claims arising out of death and disability, personal injury and loss to property and business and cover the expenses incurred by the administration (this was estimated in October 1987 at nearly Rs. 500 million in terms of the money spent on relief measures, medical treatment and ad hoc payments to gas victims).

The Indian Government said it wanted the award to be as punitive as possible so it would deter Union Carbide and other multinational corporations from disregarding the rights and safety of the citizens of those countries where they did business.

Ironically, in seeking the huge awards that only trial by jury in the United States could bring, the Government tried its best to discredit its own judicial system. Justice for the victims of Bhopal, it said, could be secured only in the United States because the 'federal judicial system...was the most appropriate forum of a just, speedy and equitable resolution of all claims'.[20] It also declared that public interest in the United States would be better served by allowing the litigation to take place there, since the company involved was an American one with units throughout North America employing thousands of Americans. The Indian Government held that a trial in the US would also focus that country's attention on the company's safety practices and other operations.

The criticism of the Indian judiciary was resented by lawyers and judges in India. Strangely enough, it was left to the accused, Union Carbide, to speak up for India's courts. Its intentions were not pure but they were clear: Indian courts were unlikely to award the kind of damages that an American judge could levy on the corporation. The courts in India were viewed as slower and, unlike in the United States, proving liability was more complex.

Judge Keenan had showed within days of the case coming to his court that he was capable of quick and fair action. He ordered Union Carbide to make an interim relief offer of $5-10 million. This would not constitute any admission of liability, he said, but would be 'a matter of fundamental human decency...if the reports I've read are true, the situation there is still critical'.[21]

Union Carbide's vice president Towe, who had returned from fruitless negotiations in New Delhi, responded with a $5 million offer.[22] He remarked that both the corporation and its subsidiary had made offers of nearly a million dollars each to the Indian Government soon after the disaster but these had been turned down. Surprisingly, India rejected the new offer as well. B.B. Singh, the Chemicals Secretary, told reporters in a 19 April brieting that he 'did not know what it was for'.[23]

The Government formally said later that the judge had placed too many conditions on the disbursing of the relief funds. Finally the funds went to the Indian Red Coss through the American Red Cross.

India's response earned the contempt of American reporters in New Delhi. 'How the hell can the Government expect an American court to take it seriously if it can't show where these funds will go?' asked Steven Weisman, the New Delhi bureau chief of the *New York Times*. 'This decision is bound to hurt the Government'.

New Delhi's stand angered the company. Warren Anderson told a shareholders meeting soon afterwards, 'it is one thing to drive a hard bargain...on the other hand, there is no way to reach a settlement without a determined effort to negotiate by all the parties.'[24] Carbide's interest in seeking a settlement, he added, should not be construed as an admission of legal liability.

Later, Anderson made another significant, point, but much more quietly. Union Carbide, he said, no longer had the capacity to meet the kind of claims that were being made on it.[25] This was very different view from statements the company had put out immediately after the disaster. Then it had said that, considering both the insurance and other resources available, the financial structure of Union Carbide would not be threatened. Stock market analysts said at the time that insurers would cushion Union Carbide to the extent of $400 million in claims. That was no longer the case, Anderson admitted, without naming a figure. The company's insurance pool had been reduced because of the likelihood of heavy compensation payments and its liability coverage, according to one report, did not extend beyond $200 million.

Carbide was not the only side interested in settling the issue out of court. Judge Keenan held a series of meetings with the lawyers' committee and Bud Holman, the company attorney, to emphasize the need for an out-of-court settlement. Otherwise, he said, the issue could drag on for years, even in the American system. After some prodding, Union Carbide and India agreed to resume talks. Parasaran and Ghosh were the negotiators for India. The negotiations were inconclusive again. To give both sides some room for manoeuvre, Judge Keennan announced the postponement of a pre-trial conference of the executive committee and

Union Carbide lawyers. The respite did not help. On 17 June 1985, the talks collapsed and Law Minister Asoke Sen summoned a series of reporters to the Indian consulate in New York to complain about the company. He repeated the charges B.B. Singh had made two months earlier and accused Union Carbide of being unrealistic and stingy.[26]

Stung by Sen's remarks, Union Carbide accused India of maligning its reputation and motives. It also elaborated, for the first time, on the sum it was prepared to pay out to Bhopal's victims. A settlement of $100 million, the company said, would pay the heirs of each dead person a hundred years of annual income in Bhopal.[27] The seriously injured would be compensated with 20 years worth of annual income.[28] The remaining funds would pay for the Government's medical and relief expenses.[29] In July, it filed its response to the Government's charges.

Union Carbide tried to portray the Indian Government as avaricious and unreasonable. 'As a moth is to the light, so is a litigation drawn to the United States.' Beginning its affidavit with that statement (taken from an earlier ruling involving an American company) Union Carbide urged the dismissal of the case in the US on grounds of *forum non-conveniens*—i.e., it held that India, not the United States, was the right place to try the case. In its arguments, the corporation pointed out that the witnesses to the disaster, the victims, the key players, the documentation and evidence, were all in Bhopal, where the UCIL plant was 'managed, operated and maintained exclusively by Indians residing in India, more than eight thousand miles away.'[30]

It said the Indian Government had passed special legislation which enabled it to represent the gas victims. The Government not only had the common law machinery but also the power and capability to deal with all Bhopal-related claims.[31] Union Carbide's counsel reminded Judge Keenan that federal courts had consistently dismissed cases brought by foreign citizens and residents involving foreign accidents because it was better they were tried in the country where the accident occurred.[32]

The company warned that if the Indian Government's petition was allowed, then the 'floodgates' to international litigation would be opened in the United States. 'The consequences of allowing such litigation to proceed in the United States (will) undermine the judicial system which permitted it.'[33]

Union Carbide said it could not establish whether the release of the gas was accidental or the result of a deliberate act.[34] This was the first reference to possible sabotage and gave some indication of the line of legal attack it proposed to pursue. It added that Sikh extremists known as the 'Black June' group (unheard of until and since then) had claimed responsibility for the disaster.[35] It was only later that it fleshed out the sabotage theory and pointed to a wholly different set of culprits.

The corporation maintained that it was India, and not UCC, which had insisted on the local production of MIC, citing foreign exchange regulations, although importing the stuff would have been cheaper.[36]

The company's affidavit also declared that no UCC director was a member of the UCIL board of directors.[37] This was correct (but headquarters did have a key man on the UCIL board—J.H. Rehfield, vice-president of its Agricultural Products Division).

Union Carbide said that Indian financial institutions owned stock in UCIL.[38] Also, the local administration largely controlled the sources of materials used in the construction and operation of the Bhopal plant[39] (the plant was built according to detailed designs provided by the Indian subsidiary of Humphreys and Glasgow, a British firm). In addition, the company reported that while the Madhya Pradesh Government had leased the land on which the UCIL plant stood to the company for 99 years it had also allowed and encouraged the growth of huts around the pesticide plant.[40]

Then it came to the heart of its argument: the complaints of liability, negligence and other charges must be decided in accordance with Indian principles of law, policy and socio-economic circumstances.[41] To substantiate its position, Union Carbide cited a 1981 judgement involving an air crash in Scotland and other rulings.

The first case that Carbide mentioned involved Piper Aircraft, a Pennsylvania firm manufacturing light airplanes.[42] When a Piper plane crashed in Scotland killing five Scots, relatives of the passengers first sued the owner, operator and pilot of the plane (who also died in the crash), all British nationals in Britain. Then the Californian administrators of the estates of the passengers filed suit against Piper and the Ohio manufacturer of the propellor, citing both firms for negligence ad strict liability.

A series of court actions resulted first in the dismissal of the case in a Pennsylvania district court on grounds of *forum non-conveniens*. But a court of appeals reversed the ruling, saying the district court had decided incorrectly. It said that such cases could not be dismissed when the alternative forum was less favourable to the plaintiff than the American forum. The case went before the US Supreme Court which rejected the opinion of the appeals court and declared that a foreign plaintiff's choice of forum deserved less deference than that of a local petitioner. Most of the litigants, the Supreme Court ruling maintained, were Scottish or English and a settlement should therefore be arrived at in the United Kingdom.

Not all such cases have been dismissed out of hand, although the benchmark ruling for litigation involving American companies overseas goes back to a case in 1947 involving Gulf Oil. The Gulf Oil judgement took into account a series of factors which would govern the issue of forum. These included both public and private interest factors: ease of access to sources, the cost of arranging witnesses' depositions, the possibility of a view of the premises and other factors which could make any such trial easy and inexpensive. Other factors mentioned were administrative difficulties flowing from court congestion, the local interest in settling local controversies locally and the need to have the trial in a forum that was conversant with local laws.[43]

However, American courts ruled differently in a 1975 case involving Vietnamese children who were injured in an accident during a babylift during the last days of the American raj in South Vietnam.[44] The accident occurred after the door on a Lockheed transport aircraft blew out and the plane crashlanded in a rice field near Saigon.

Lockheed's motion for dismissal of the case on *forum non-conveniens* grounds was rejected because the court felt there were no alternative forums in Vietnam.[45] It pointed out that the evacuation was an American one, organized and implemented by Americans.[46] Besides, all the witnesses were in the United States, including the nurses who had tended the children and the crew of the flight.[47]

In another judgement, a US court reversed a district court's ruling dismissing a case against Avon Products, an American cosmetics firm, for alleged fraud involving commercial activities in Taiwan.[48] It declared that the central principle of the Gulf Oil doctrine was that unless the balance was strongly in favour of the defendant, the plaintiff's choice of forum should rarely be disturbed.[49] It pointed out that Avon Products was a New York-based firm.[50]

In its affidavit before Judge Keenan, Union Carbide asserted that the New York court lacked the insight and background which only an Indian judge could bring to bear on such a case.[51] It would surely be unfair to apply the American approach to US corporations owning stock in foreign companies if US 'values' were not applicable to all firms operating under foreign jurisdiction.

The corporation then pointed out that a chemical leak elsewhere in India in June 1985 had injured 113 persons.[52] This was a reference to a chlorine leak from a textile firm in Bombay where compensation payments had been far less than that sought by India against Carbide. 'Should the Bhopal claimants be treated ...differently just because a US corporation owns 50.9 per cent of the stock of the company which owned and operated the plant?' the Carbide plea asked.[53] Another point the company brought up in favour of transferring the case to India was that the intensive investigation necessary to determine whether the disaster was the result of an accident or sabotage could only be conducted in India.[54] Finally, the company maintained that a trial in the United States would be expensive and an administrative (one problem mentioned here was the difficulty of translating statements of witnesses from Hindi into English) and judicial nightmare.[55]

Carbide's presentation was firm and polished. To support its theme—that the Indian judicial system was adequate and competent—the multinational produced one of its most eloquent performers, Nani A. Palkhivala, one of India's best known lawyers and a former ambassador to Washington. He was principled, one of the country's best legal minds and a prize catch for Union Carbide.

'I have done the state some service'; with that ringing quotation from *Othello,* Palkhivala launched a brief but impassioned defence of India's judicial system and Union Carbide's demand for a trial there.[56] He declared that Indian courts were competent, innovative and acted quickly on public interest cases. 'India's Supreme Court, in the range of its power and the sweep of its jurisdiction, is without a rival in human history. Its writ runs over more than one-seventh of the human race.'[57] His exposition was interspersed with references to his own record in procuring justice for his clients on complex legal and constitutional issues.

'I am constrained to say that it is gratuitous denigration to call the Indian system deficient or inadequate,' Palkhivala intoned.[58] He held that the only possible inadequacy in the system was its inability to award the kind of damages that American juries could hand down'. He also referred to the gas leak from the Shriram Food and Fertilisers company in Delhi, almost exactly one year after the Bhopal tragedy, which had injured scores and created a panic across the city. Palkhivala noted that the Supreme Court had ordered its own inquiry into the leak within two days of the disaster.[59]

'Is there any reason to assume that the courts would not act expeditiously for thousands of victims of the Bhopal tragedy, when they did so for far fewer sufferers of the gas leak in Delhi?' he asked. [60] The Palkhivala presentation was one of the turning points of the case: indeed, Judge Keenan cited it in his judgement.

The Palkhivala testimony came in response to a scholarly exposition by Marc Galanter, a professor of law and South Asian Studies at the University of North Wisconsin. He was regarded as an authority on Indian law and his books had even been cited in Supreme Court judgements in India. Galanter was picked by the Indian Government to portray a creaking judicial system, overburdened by delays and choking under the pressure of tens of thousands of untried and unresolved cases.

He said that civil litigation had remained undeveloped in India because the Supreme Court had been occupied for many years with constitutional matters.[61] He pointed out that India had only 10.5 judges per one million population, one-tenth of the proportion of judges to people in the United States.[62] Galanter quoted extensively from the Indian Government's own findings. He also reviewed the vast number of pending cases in the Indian Supreme Court alone: a total of 46,074 cases were launched in 1984. Less than 36,000 cases were decided that year, leaving a backlog (growing from earlier years) of 86,733.[63]

Galanter added that a study of tort cases in the Indian high courts and the Supreme Court from 1975-84 had turned up only 416 cases, 360 of which were related to motor vehicle offences.[64]

Of the fifty-six others, he said, twenty-two related to claims involving negligence or liability.[65] These included railway crossing accidents, the death of a child on a school picnic (the teacher was blamed for the death), medical malpractice suits and cases involving embezzlement of bank funds. The bar in India lacked the skills, the experience, the degree of specialization and the capacity to respond to the demands of mass disaster litigation, the expert said.[66]

He also pointed out that discovery, the examination of documents or witnesses by contesting parties, was strictly limited in India. 'The rules allow interrogations, inspection and production of documents and material evidence. But there is no provision for pre-trial depositions and they are unknown to Indian practice', Galanter said.[67] Pre-trial depositions are a feature of American courts which allow the examination of witnesses in discovery proceedings. This is not allowed in India. The American format of pre-trial conferences enabling both litigants and the judge to hammer out courtroom procedure and the consolidation of cases is also unknown.

Discovery proceedings are often stalled in India when one party declares it has no documents relating to the matter. The party cannot be pressured into revealing whatever documents it does have. As a result of this, Galanter said, discovery and inspection of documents was one of the 'least used parts of the Indian Code for Civil Procedure'.[68]

Later, Judge Keenan was to address the limited role of discovery in his judgement and rule favourably for India when he sought to bind Union Carbide to the rules of discovery effective in American federal courts.

A memorandum filed jointly by the Indian Government and the attorneys for private plaintiffs urgd Judge Keenan to toss out Union Carbide's motion of inconvenient forum.[69] Replying to the company affidavit, the lawyers reiterated their contention that Indian courts were inadequate.[70] They said that nowhere had the corporation even offered to abide by an Indian verdict, although it had been praising Indian courts.[71]

There were other points mentioned in the joint memorandum filed by the Indian Government and private plaintiffs. One was that by trying to push responsibility for the disaster on UCIL, the parent company was actually seeking to limit claims to the capacity of the Indian subsidiary.[72] UCIL's assets in India were worth less than $100 million, the memorandum stated, and would be grossly inadequate to meet compensation demands.[73] The memorandum also said that design or product defect alone had been held adequate to dismiss a plea for *forum non-conveniens.*[74]

'Union Carbide was keenly aware, through its safety audits and monitoring of the plant, of the obvious potential for disaster at Bhopal and had ample opportunity to take corrective action', the memorandum said.[75] 'This factor alone distinguishes the Bhopal litigation from the majority of forum decisions granting dismissals and is a compelling reason for exercising jurisdiction'.[76] It said that crucial evidence was to be found in the United States. Referring to Union Carbide's praise of the Indian courts, the lawyers' committee concluded: '(Union Carbide) has no standing to assert this belated concern for the welfare of the Indian populace. The victims and their Government have spoken by the filing of their complaints in this court'.[77]

•

In the middle of the legal fight, Union Carbide also fought off and survived a costly and exhausting bid to buy it out by GAF Corporation, a small chemical firm based in New Jersey. The takeover plan was announced by Samuel J. Heyman, the GAF chairman, who declared an offer of $68 per Union Carbide share in December 1985.[78] His company already owned ten per cent of the multinational's total stock and wanted to increase it to a whopping ninety per cent. Payment for the shares would be in cash.

A series of offers and counter-bids for equity followed. In January 1986, Union Carbide made a last ditch, desperate effort to save itself from a takeover. It mortgaged its assets to the banks so that it could buy back fifty-six per cent of stocks from its shareholders and defeat GAF.[79]

It was a very expensive business. To borrow $1.1 billion to buy back the shares, Union Carbide had to swallow its pride, take a hard look at its extensive business operations and decide areas of priority investment, for the latest borrowings would add another $350 million to the interest rates the company would have to pay every year (it already paid $200 million of interest expenses on existing debts).[80]

UCC eventually decided to cut back on non-strategic businesses. The 'strategic' businesses which it was to retain included plastics and chemicals, industrial gases, carbon products and speciality business and services.[81] Various other product sections went to interested bidders. For instance, Eveready, the battery division, went to Ralston Purina.[82]

Following Union Carbide's steps to thwart the takeover, GAF and three other companies contracted an agreement with Union Carbide to freeze their stock holdings at 10.6 per cent of all Carbide stock for ten years.[83] The accord followed eighteen months of battle in boardrooms, stockholders meetings and courtrooms.

But the bruised company still had to think of ways of raising funds to meet its debts. It had already decided to sell off businesses which were not considered crucial to future operations. But that wouldn't meeet the enormity of its debt. So it decided to sell property: its Danbury headquarters went for $340 million to a real estate developer.[84] Having sold the property, UCC leased back its buildings for twenty years.[85] This sort of sale lease arrangement is what other large American corporations are doing in the United States in an effort to beat new high taxes that came into being after December 1986. These taxes are aimed at major capital gains transactions.

The December 1986 date is also important because that is when the company's elaborate restructuring plan finally began moving. That too was designed to beat the tax rap. Union Carbide's problems were viewed unsympathetically in India. Indian journalists and activists accused the Indian Government of silently standing by, allowing the corporation to strip itself of all major assets and holdings so that it could not meet any eventual ruling on compensation. This charge wasn't wholly correct for in December 1985, during the height of the Carbide-GAF battle, India's counsel, Michael Ciresi, dispatched a notice to UCC's board of directors saying that India was dismayed by Carbide's plans to buy back shares from its own stockholders.[86]

'The financial impact of the company offer on Union Carbide will be devastating', the notice said. 'The company's net worth will be decimated, its ability to borrow eliminated, and its ability to meet its obligations as they become due severely curtailed'.[87]

Ciresi added that India's claims for compensation took precedence over any debts the corporation might have accumulated and that the Indian Government would seek to block any moves that would endanger its demands.[88]

'More than two-and-a-half-billion dollars of the company's assets have already been sold since the corporation's "restructuring programme" was announced in August of 1985,' Ciresi maintained.[89] He declared that payments to stockholders from such sales would be unlawful 'if they are made at a time when the corporation is or will be rendered insolvent'.[90]

Ciresi threatened legal action to prevent Carbide from going ahead with the liquidation of its businesses because it could hurt the compensation claims.[91]

Barely three weeks later, came Judge Keenan's verdict bundling the case back to Bhopal. And it was not until November 1986, through an injunction passed by the Bhopal District Court, that India won a pledge from Unión Carbide that it would keep assets worth $3 billion to meet any future verdict on compensation.

Through the summer and fall of 1985, Judge Keenan instituted discovery hearings with both sides furnishing documents and information.

The judge appointed a special magistrate to listen to each side, giving each team a total of twenty days for document inspections and questioning. The hearings continued away from the fuss of publicity until the end of that October. By this time, Judge Keenan had also imposed a code of confidentiality on those involved in the negotiations. They should not talk to the press because of the sensitivity of the matter, he said. News reports would only hurt the process and destroy chances of an out-of-court settlement. The three-man committee, appointed by Judge Keenan and Union

Carbide, agreed to the conditions.

In December 1985, Union Carbide's attorney tried to get the sabotage theory accepted.

There was a presentation too by Robert Hagler, a lawyer who represented more than 20 public interest groups which formed a coalition called the Citizens Commission on Bhopal. Hagler's stand resembled New Delhi's. Perhaps his perception and that of other activist groups in the United States was subconsciously influenced by the knowledge that the suffering at Bhopal was caused by an American firm. Hence their tone of moral outrage and condemnation was stronger.

Hagler based his argument on two precedents: the Dalkon Shield case in which Robins, Zelle had established that the manufacturer was liable for a defective birth control device.[92] The second case involved the Manville Corporation, the largest American manufacturer of asbestos, which filed for bankruptcy after being hit for billions of dollars in health damages. The victims won a major victory when the company agreed to transfer fifty per cent of its stocks to a $2.5 billion trust fund for the plaintiffs in the spring of 1986.[93]

Urging an American trial, Hagler asked the court to create legal history by settling a mandatory code of conduct for transnational corporations.[94] 'United States courts are the most experienced in the world in handling tort litigation of this kind, arising from the worst industrial disaster in history', the Citizens Commission's brief said.[95] 'This case provides an opportunity to bind all American multinationals to a standard of care that will prevent the recurrence of a catastrophe such as Bhopal.'[96]

The report added that if the court 'declined the challenge', American corporations would continue to practise double standards.[97]

In January 1986, the judge reserved his verdict on the hearings, saying he wanted to meet witnesses from Bhopal, including doctors, orphans and officials. Quickly, the Madhya Pradesh Government bundled together a team of eight persons and flew them to New York. Among the Bhopal victims in the group were a railway engineer, a salesman and a fifteen-year-old orphan who had lost seven members of his family. The officials accompanying the victims included Ishwar Dass, the relief commissioner, Deep

Chand Yadav, the man who succeeded Bisaria as Mayor of Bhopal, and two doctors, N.C. Mishra and Harihar Trivedi of the Hamidia.

For two days, the judge met the survivors and the medical men. At most of the meetings, the Indians were accompanied by Dass. The whole visit was shrouded in secrecy and Indian reporters in New York were turned away by embassy officials who cited Judge Keenan's order of confidentiality. Yadav was handed a letter by an Indian consular official as soon as he arrived, instructing him to stay clear of the press.

About the same time, the judge also met separately with Warren Anderson, the Union Carbide chief, and Parasaran, the Indian Attorney-General. At his suggestion, the American and the Indian met secretly in February 1986.[98] Anderson was still keen on a settlement and offered to hike the Carbide offer to $300 million.[99] The Indians were not prepared to touch the bait.[100]

Meanwhile, Judge Keenan talked to Mishra and Trivedi, the doctors. Both explained that the health of the majority of the victims continued to be poor. Mishra says he spent four hours with the judge who indicated that he felt the Madhya Pradesh Government had not done enough at the time of the disaster for its people. Among other things he referred to the failure of the Government to evacuate Bhopalites after the gas spread. 'He was not convinced of the legal responsibility of Union Carbide although he was convinced that it had a moral responsibillity', Mishra says.[101]

The doctor says the judge told him that the corporation had even offered a modern 1,000-bed hospital and pledged to support it financially for 30 years as part of a compensation plan.[102] The Indian Government had rejected the suggestion.

When the Indian Government did not respond to the Anderson offer, Union Carbide went to the lawyers representing the private lawsuits. Would they accept $300 million in a trust fund to compensate the victims? The lawyers were quick off the mark. They agreed to discuss the issue among themselves and then with the company's lawyers. It seemed the best they could get, especially as the case could be sent back to India, and with it their hopes of big fees and more publicity. So they accepted the offer and on 19 March 1986 agreed to pursue a full settlement.

The report of the $300 million deal was first published in the *New York Times.*[103]

Within days of the *Times* report, John Coale and Arthur Lowy, the first American lawyers to land in Bhopal in 1984, returned to India. They tried to sign up fresh clients and get a power-of-attorney from others. Take the $300 million, they told Bhopal's victims. It was the best they could expect from Union Carbide.

The welcome this time was cool, even hostile from some groups. Policemen followed the lawyers everywhere and questioned those who met them. Coale later claimed his company had signed up 800 persons and that the public response to his statements had been good. He accused the Madhya Pradesh Government of spreading 'vicious lies' against him. While all this was happening, the Indian Government issued a strong repudiation of the settlement. It was not involved in any deal, the Government said, and pointed out that an out-of-court settlement would be infructuous without its concurrence and that of the judge. Judge Keenan disapproved of the publicity and Carbide's moves. He called an emergency meeting of the litigation committee and Carbide's lawyers. He wanted to know who had broken the rule of confidentiality. No one was prepared to own up. In frustration, he told the lawyers that he was stopping his efforts at an out-of-court settlement. 'You've broken the confidentiality that was binding on this court; from now onward you can take public positions, there will be no secret talks', he was quoted as saying by one of those present at the meeting.[104]

The Keenan verdict came on 12 May. It was an interesting judgement, well thought-out, far-reaching in its implications, and well phrased. The pivot of the judgement read: 'No American interest in the outcome of this litigation outweighs the interest in India in applying Indian law and Indian values to the task of resolving this case'. Referring to Indian arguments for a trial in the United States, because the courts there were not capable of conducting the cases arising out of Bhopal, Judge Keenan said that this was not sustained by any evidence presented by the Government. He paid tribute to the Indian system, saying it was not right for an American court to retain the litigation. 'This would amount to imperialism. The Union of India is a world power in 1986, and its courts have the proven capacity to mete out fair and equal justice.' He upheld the Carbide plea on the choice of forum, saying that India was the best place for the trial, because of the presence of witnesses and documentary evidence. He said that Palkhivala and J.B. Dadachanji, another Supreme Court lawyer and Union Carbide counsel in India, had presented persuasive evidence on the efficacy of the Indian system.

Although he accepted the plea to return the cases to India, the American judge imposed three conditions on Carbide. These conditions, the company was told, would be binding on it. If it refused to accept them, the case would remain in the United States. These were:

1. Union Carbide shall consent to submit to the jurisdiction of the courts of India, and shall continue to waive defences based upon the statute of limitations.
2. Union Carbide shall agree to satisfy any judgement rendered by an Indian court, and if applicable, upheld by an appellate court in that country, where such judgement and affirmance comport with the minimal requirements of due process.
3. Union Carbide shall be subject to discovery under the model of the United States federal rules and civil procedure after appropriate demand by the plaintiffs.

In essence, Judge Keenan gave both Union Carbide and India their chief demands. The central Indian plea was that Union Carbide, the parent firm, and not UCIL, be recognized as the chief accused (its subsidiary was worthless to the Government, it had assets worth only a fraction of the parent company). The court order gave the Government access to Carbide's documents in the US and in India and also made the court orders binding on the multinational. It wasn't exactly what the Indian Government wanted but it represented more than Carbide had originally been prepared to give.

The lawyers handling the private lawsuits decided to appeal the judgement in a superior appeals court. After days of study, Union Carbide announced that the judge's conditions were not wholly acceptable. It objected to the last one which gave the Indian Government access to Union Carbide documents in the United States and elsewhere, but did not give the corporation a chance to examine official records. This was neither fair nor proper, Carbide's attorneys said.

The lawyers for India had told the judge earlier that they were prepared to be bound by US federal rules on civil procedure governing discovery rules imposed on Carbide. In fact, Ciresi says he even wrote the company a letter on 23 May suggesting that both sides initiate discovery procedings in India to the extent that New Delhi itself used such facilities.[105] Expanding on the subject, Ciresi declared that the Indian Government would not seek discovery with respect to individuals or information customarily located inside India without agreeing to submit to similar discovery by Union Carbide Corporation.[106]

The multinational was not satisfied. It wanted judicial confirmation. So, in July, Union Carbide appealed this particular condition of the Keenan judgement before the US Court of Appeals for the Second Circuit in New York.

Union Carbide was making an important point in its appeal: by allowing the Indian Government use of American discovery procedures, it said, Judge Keenan was being unfair for he had not given the same facilities to the corporation. The Government's offer to allow partial inspection of documents in its possession (relating to liability and the accident) was seen as inadequate. For not only was New Delhi to have unrestricted access to Carbide's inner secrets and strategies but it was appointing itself the arbiter of how much Carbide could see of the Government's material. That was clearly going to hurt Carbide and the corporation would have been foolish not to contest the condition. Carbide won a positive ruling in early 1987 on this issue although the appeals court reaffirmed the decision to send the case back to India. The ruling of the appeals court was further appealed by India to the US Supreme Court.

It declined to overturn the verdict of the appeals court. The Indian Government had lost in all three American judicial forums.

But to return to the events of 1986. As the legal wrangling continued in the US, the Indian Government shifted its case to home soil.

Chief Justice P.N. Bhagwati and Law Minister Asoke Sen were

cautious in their initial reactions to the Keenan judgement. They needed to read it fully, they said, although they liked what newspaper reports said of it. Sen particularly spoke of the conditionality on Union Carbide. When he returned to New Delhi, Sen went into meetings with aides to consider a quick way of tackling the issues raised by the judgement. Various options were discussed: hearings in the Delhi High Court with an appeal to the Supreme Court; hearings in Madhya Pradesh, with the cases consolidated under a high court judge; an exclusive court in Bhopal at the level of the district court there, where 6,000 cases had alreay been consolidated (legal action in the cases in the district court had been stayed in 1984, when the Government said its proceedings could hurt the legal battle in the United States). The last alternative, that of the district court, was selected for what the Government hoped would be a swift trial.

Swift, despite Palkhivala's glowing testimonial, is what the country's courts are not.

There are tens of thousands of cases pending before the small, medium and higher courts across the nation. These include land disputes, cases of brides being burned because they brought too little dowry, religious litigation, cases of caste oppression, rape, robbery, espionage, murder and accident compensation. The clogging of the system has meant that an estimated 87,000 people were in jail 'illegally' in 1984,[107] often for petty offences, without their cases even coming up for trial (often such people leave prison as hardened criminals, embittered by the system. Others who are eventually released by public interest litigation after decades of imprisonment just do not know how to adjust to what appears to them a frightening and strange, new world). Many cases come to trial, even in the Supreme Court, after the litigant has died. In 1986, the country's highest court was deciding cases which began in 1974 and earlier.[108] The Chief Justice was reported as saying 'the court is on the brink of collapse'.[109] According to M.J. Antony of the *Indian Express,* 'more than half the time of the (Supreme) Court is taken up in mere admissions and the court has become a court of appeal and interim orders with hardly any time to give final judgements'.[110]

It must be said here that the Indian Supreme Court has acted swiftly on important constitutional, environmental and human rights issues, rebuking the Government, chastising jail officials and ordering the redress of decades of injustice. But such instances are not numerous.

Obviously the Bhopal case would need to be resolved swiftly. But delays appeared inevitable. For one thing, the claims directorate at Bhopal was taking a long time to complete its scrutiny of the 518,000 compensation claims.[111] There was the question of fraudulent claims: over 700 had been identified in 1986 alone and two persons had been arrested for filing such claims and getting relief funds from the Government.[112] The Government's assessors needed to decided the amount of compensation to be paid to different categories of victims: to relatives of the dead, the chronically ill and those who had suffered minor injuries. It was only towards the end of 1987 that this aspect of the litigation was speeded up.

Naturally enough, the gas victims were depressed at the rate of progress. Ladli Saran Singh, a lawyer who filed 1,000 cases on behalf of nearly 10,000 gas victims in the Bhopal court, felt the Government wasted more than a year by seeking compensation in America.[113] 'The courts have delays here but justice is done', he said.

On 5 September 1986, nearly four months to the day that Judge Keenan had dismissed the Indian plea, the Government of India flew Shyamal Ghosh, of the Chemicals Ministry, to Bhopal to file its compensation case against Union Carbide. The plaint reflected the lack of imagination on the part of the Government, for apart from a few paragraphs, the complaint was virtually an exact copy of that filed by Robins, Zelle before the New York court.

Again, the Government said it had no exact casualty figures, although it mentioned a provisional death toll of 1,700. This figure was two years old. Even the Relief Commissioner at Bhopal had a higher figure of 2,100 in July 1986. Obviously, one department of the Government did not know what the other was doing. (By the end of 1986, the Government death toll read 2,400, with another 30,000 to 40,000 seriously injured. In 1987, the figure was 2,850 dead.)

The September complaint reflected a sad lack of official will to move speedily and coherently on the disaster. It was not until two months later that the Government declared the amount of compensation it sought from UCC.

After more delays, this time as a result of a series of adjournments by Carbide lawyers who said their clients in Danbury had not received a summons for the case, the legal battle was joined in earnest in October 1986 in Bhopal before Judge G.S. Patel.

It was at this point that the Indian Government filed an affidavit urging the judge to stop Carbide from selling its properties, as it would restrict its ability to meet any damages awarded to the Bhopal victims. Carbide, of course, disagreed strongly with the Indian Government's contention.

After hearing both sides in mid-November, Judge Patel declared that he was restraining Union Carbide temporarily from selling its property and rebuying debentures related to its plan to liquidate debts incurred in the GAF takeover bid. Though the company was furious, it look comfort in the fact that it had supplied the court detailed affidavits which sketched out how its $2.5 billion recapitalization plan was aimed at strengthening its financial condition.[114] It said the refinancing plan was designed to reduce its debts and interest expense, while improving earnings and cash flow.[115]

But more interestingly, Carbide's affidavit denounced the Central and State Governments for their alleged liability for the disaster.[116] The affidavit said they knew at all times of the dangers from MIC and other toxic gases and chemicals used at the Bhopal plant,[117] that New Delhi had okayed a waiver of responsibility by Carbide for any mishap that occurred during the operation of its UCIL factory, and that by allowing illegal slums to grow up around the plant (and then legalizing their existence), the State Government was liable for the high death toll.

The 169-page document gave the fullest and most cogent view the company had hitherto revealed on the disaster and why it could not be held culpable. In fact, the affidavit denied every charge levelled against Carbide including the fact that it had either operated or controlled the plant, because it claimed the factory was run by UCIL. UCIL, it said, was a separate company.

It expanded on the sabotage theory (without naming the alleged saboteur) and also alleged that there was a conspiracy between workers and official investigators to conceal evidence and the guilty. It said very clearly, too, that the Government had played a key role in the functioning of the plant: as many as twenty-two Central Government agencies and ten State Government agencies had been associated with it.[118] These departments included the Chemicals Ministry, the Directorate-General of Technical Development, the Controller of Imports and Exports, the Plant Protection Adviser and the Central Insecticides Board.[119] On the state side, the agencies involved included the Chief Inspector of Factories, the Water Pollution Prevention Board, the Directorate of Industrial Health and Safety and the Departments of Agriculture, Commerce and Industries.[120]

Carbide's affidavit also declared that as far back as July 1975, nine-and-a-half years before the accident, and four years before the plant's start-up, Madhya Pradesh's Chief Inspector of Factories had 'asked for and received details of the properties and hazards' of the chemicals to be used at the Bhopal plant.[121] Included in the information package was material on chloroform, MIC, phosgene, carbon monoxide, chlorine, hydrogen chloride (hydrochloric acid), carbon tetrachloride, napthalene, dichlorobenzene, chlorosulfonic acid, monomethylamine, and sodium hydroxide.

The material it had provided, Carbide said, warned of the hazards of exposure to MIC and said the compound was a reactive, toxic, volatile and flammable liquid.[122] Yet nowhere in its statement, did the corporation say it had recommended an antidote or line of treatment for MIC victims at the time of this declaration.

It said that during the curfew ordered at Bhopal, as in numerous other North Indian cities during the anti-Sikh riots which followed the assassination of Indira Gandhi in 1984, UCIL sought special passes for its workers to keep the plant operational through the trouble. Its reasoning was that if the plant was left unattended, an emergency might arise which could endanger the safety not only of the plant but also of the surrounding area.

The company's repeated assertions of the plant's dangers, the fact that the Madhya Pradesh Government gave as many as seventy-eight approvals for different kinds of construction work, and the point that the Chief Inspector of Factories, who also performed a double role as Deputy-Director for Industrial Health and Safety, had cleared the plant at an inspection two weeks before the disaster, all showed that the Government knew—however vaguely—of the hazards. These also were the reasons that figured in Carbide's countersuit against the Madhya Pradesh and Central Governments in which it said that its opponents should be ordered to pay to Union Carbide a share of the damages proportionate to their responsibilities and liabilities.

The corporation said in its suit that it was UCIL which controlled the construction of the plant. In other words, it was cutting itself totally off from its subsidiary, although it could not hide the fact that it was a majority stockholder in UCIL. Carbide also referred to an agreement between the two companies (UCC & UCIL) duly approved by the Centre which had a clause saying UCC, which had submitted an exhaustive design package for the plant's setting up, 'would not in any way be liable for any loss, damage, personal injury or death resulting from or arising out of the use by UCIL of the design packages'[123].

Nothing could have been more categorical. But officials in the Chemical Ministry say there are thousands of such technical collaboration agreements in the country. 'Government approval of an inter-corporation accord is routine and can hardly be described as a waiver of liability', says one official.

Interestingly UCC alleged that the Indian Government had insisted it (UCIL) use local materials (which could have been substandard) for plant construction. However, it must be clarified here that UCC had undertaken to provide, from Danbury, all specifications of the materials to be used in the construction of the plant.

Another issue on which the company faulted the Government was its reluctance to allow American technicians to continue for long spells at the plant (this springs from the Government's desire to exercise control over foreign-controlled subsidiaries, ensure local technical expertise develops and restrict the outflow of foreign exchange).

Union Carbide acknowledged that it had trained twenty UCIL

employees in 1978 and 1979 in the United States in very extensive programmes which included safety, operations and operating procedures, process and personnel safety, environmental protection, toxicology, and administration. A UCIL medical officer was singled out for special mention. The company said he had received 'extensive training' in safety, especially emergency treatment and toxicology.

Apparently, the training programmes were not adequate for in a March 1980 letter to the Ministry of Petroleum, Chemicals and Fertilizers, UCIL said its technicians were not competent to run the Bhopal plant without the supervision and assistance of an American specialist for a minimum of two years.

The Government allowed an American specialist, Warren Wommer, to stay until May 1981, then extended his stay permit till August 1982 and then again until December 1982, a period of two-and-a-half years. Former employees described Wommer as the man who effectively ran things at the plant, although he was supposed to report to the general works manager. The American returned to the United States after the Government declined to permit further renewals of his contract.

The Indian Government was enraged by the corporation's suit. In November 1986 it specified its compensation demand for the first time: Carbide would have to pay a minimum of $3.1 billion to the 2,850 who had died and those still suffering long-term effects of the gas.

The compensation figure was daunting and three times that which Delhi had sought in the 1985 out-of-court settlement talks. 'A lot of time has passed since then, a lot more evidence of damage and death has emerged, there have been more expenses by the Government', said one Indian official.

Finally, after long consultations with Danbury, John A. Clerico, a corporation vice president and treasurer, filed a special affidavit before the Bhopal court saying Carbide was prepared to pledge a minimum of $3 billion in assets in reserve to meet any eventual decree against it. He offered to furnish a quarterly certificate verifying the existence of such assets by an independent valuer, without acknowledging liability for the disaster. The Government accepted the offer and Judge Patel held Carbide to it. Soon after this Judge Patel stepped down from the case when it was discovered that he was a claimant himself and Judge Deo took over.

# Who Was Responsible?

In their own way, to differing degrees, each of the four major participants in the Bhopal drama—Union Carbide, UCIL, the Governments of India and Madhya Pradesh—was responsible for the disaster and its aftermath.

There is a moral and cultural responsibility that goes beyond legal hairsplitting, beyond judicial rulings of liability, beyond profits and public posturing. And it is certainly true that in their response to the Bhopal disaster there was a lack of honesty on the part of the principal protagonists that deepened and lengthened the pain of the victims. The delays caused by truculent antagonists, the failure of local bodies to respond adequately to the disaster (except for the initial superb burst), the consistent pettiness displayed by Carbide, all these were avoidable and totally unnecessary. Instead of learning from each other and their mistakes, the principal characters in the drama were more interested in defending themselves. This was unfortunate, for legal culpability can never be the only yardstick by which responsibility for tragedy can be gauged.

Thus Union Carbide was wrong to dispute its control over the Indian sector of its worldwide empire. The multinational's American officials surveyed and selected Bhopal for the MIC plant. And they set up a plant that was badly flawed from the beginning and a losilng proposition. It lost Rs.20-40 million every year.

It was as R. Natarajan, a director at Union Carbide Eastern said, 'an oversized plant with an undersized market'[1]. By the end of 1984, Union Carbide had decided to dismantle the units at Bhopal and export them to Brazil and Indonesia. But UCIL warned headquarters that the MIC unit's dismantling and shipment would be a problem 'because of the high corrosion at several points and some tall columns, which may need some work at the other end.'[2]

In their own unambiguous language, the managers of Carbide were emphasizing their duplicity. On the one hand, they claimed the plant was in good shape. But on the other, they admitted—especially to their bosses—that things were seriously wrong and needed urgent repairs.

The company did not obviously give as much attention to workers' training as it should have. Nor did it share its information updates on safety as promptly with its subsidiary as its should have done.

From the beginning of the case, Union Carbide sought to distance itself from its Indian division instead of accepting even a degree of responsibility.

This was patently wrong. Key personnel training was done in the United States. Jagannath Mukund, the works manager at the time of the mishap, had spent three years at Institute before being transferred to Bhopal. Gokhale, the managing director and now chairman, was based at New York between 1973 and 1975. The two men were among a regular flow of Carbide personnel between India and the United States. Letters, telephone calls and telexes kept Danbury and Hong Kong well informed. UCIL was run by an American managing director until 1978 when the first Indian chief executive took charge.

The main safety audit of the Bhopal plant was done in 1982 by the Poulson team out of Danbury. It located ten major safety flaws. Yet, there was no independent survey between then and the accident to check on Mukund's claims to headquarters that all possible dangers had been tackled.

The design plans for three critical systems that failed—the gas scrubber, the flare tower and the water spray system—came from the parent company. The technical manuals for Bhopal were based on original documents issued by Union Carbide Corporation.

The management of the company was supervised by Union Carbide Eastern. Four Union Carbide Eastern executives were on the Indian firm's board of directors. All of them had earlier worked for Union Carbide. J.M. Rehfield, an executive vice president at Danbury, was a member of both boards and had resoponsibility for Asia.

The technology used to manufacture MIC was American: it was the same as that used at Institute. A former managing director at UCIL says that the American managers overrode objections to the storage of MIC in large tanks, saying that any other form of storage would not be economical.

The American firm held a majority ownership of the company. No major decisions could be taken without its consent and consultation. When UCIL wanted to sell its Bombay plastics unit in 1983, the decision had to be cleared by headquarters. Its annual budget had to be cleared by UCC as well.

The fact that such close links existed between UCIL and the parent company UCC made the buck passing that ensued post Bhopal all the more unfortunate.

Other aspects of UCC's dealing with its subsidiary are questionable as well. Take, for instance, the handling of a report of a 1984 safety audit at UCC's Institute MIC II Unit.[3] The report said that a major concern among company technicians was a possible runaway reaction in the MIC storage tanks.[4] The report was sent to Bhopal only after the accident, although it detailed the inherent dangers of manufacturing MIC. The failure to send the report is inexplicable considering that both UCC chief executive Anderson and Warren Wommer, a manager who had worked at UCIL, claimed at the time of the accident that the Institute and Bhopal plants were similiar. However, it must be said here that Mukund, who saw the report in February 1985, says he is not sure whether an advance look at the report would have made any difference. For one thing the cooling systems for the MIC tanks in the two plants were different. 'But the truthful answer is we do not know if it would have prevented the tragedy', Mukund says.

It is instructive to take a detailed look at the Institute safety audit. The audit was dispatched by Poulson, who had conducted the Bhopal safety inspection two years earlier, to B.D. Mollowan, the deputy manager at the Institute plant.[5]

One major concern expressed by the six-member team of safety experts was the potential for a runaway reaction in unit storage tanks due to a combination of several factors and reduced surveillance.[6] A second worry was the chance that MIC plant workers could be seriously overexposed to chloroform, a by-product of the reaction.[7]

Marked 'Business Confidential', the report speaks of the various possibilities that could cause a runaway reaction. One was the non-continuous use of units (where the tanks are used for long-term storage, compared to their rather transient operation when the unit is running). A result of this would be that 'the tanks tend to get less attention and be sampled less frequently with the higher probability of contamination going undetected for a relatively long period of time'.[8]

Water contamination, which can take place over a period of time, could also cause a runaway reaction, the report warns.[9] While some cases of water contamination may have been handled with little or no problem, the Poulson report remarks that complacency and lack of concern could allow such a situation to deteriorate.[10] 'This combination of water and catalyst contamination possibilities, reduced surveillance, increased residence time, and an experience-based low level of concern toward the potential hazard leads the team to conclude that a real potential for a serious incident exists'.[11] The committee sums up the report with the suggestion that tanks be sampled at least daily or after every cooling cycle while idle. This was not being done at Bhopal, increasing the chances of contamination.

Apart from the Institute safety report, another pertinent piece of documentary evidence which Carbide now does not choose to bring up is the detailed first report the corporation put together of the disaster. This was presented in March 1985. UCC faulted the Indian subsidiary on six major violations of company safety rules.[12] No less a person than Warren Anderson said, 'The plant should not have been working without procedures being followed'. But it is justifiable to ask: Who laid down the procedures? The answer to that is UCC, although Anderson qualified his remark by saying that safety was a 'local issue' and UCIL's responsibility. And despite the corporation's contention that Mukund was a UCIL employee at the time of the accident there can be no gain saying the fact that he was trained in every aspect of the handling and running of an MIC plant at Institute.

But reports apart, it is inalienable that a number of operating procedures were regularly violated at the Bhopal plant: well-qualified workers were replaced by underqualified technicians; the plant was understaffed; the number of MIC workers on each shift dropped from fifteen to eight. And finally, the question of the saboteur. As we've seen earlier, Carbide's theory goes that a disaffected worker let water into an MIC tank and it leaked. But this is essentially a simplistic explanation. It does not explain the lack of crisis preparedness on the part of the other workers (who were not 'saboteurs'), it does not brush away the malfunctioning or ineffectiveness of key accident inhibitors, it does not exculpate the company on points of inadequate safeguards and warning systems. On all these points Carbide cannot avoid its responsibility.

What of government responsibility? Neither New Delhi nor Bhopal can avoid the charges against them, however hard they try. Perhaps the local Madhya Pradesh administration can be held *more* responsible on the following counts:

* In 1982 the state's Labour Minister declared the plant safe even after a worker had been killed in a phosgene leak. He had no need to make such a categoric statement.
* Three major accidents which injured dozens of workers occurred within months of the minister's statement, showing it to be ill advised and incorrect.
* The inquiry into the phosgene death began in 1982 and was submitted in 1984. It stayed for seven months in a dusty office in the Labour Department without being studied. In October 1984, it went to the Labour Commissioner, who took neither notice nor action on the report until after the disaster. He was one of six officials who were suspended, sacked or sent on leave after the leak. The local Government was thus not unaware, as it has been saying all along, of the lethal properties of MIC and other gases at the plant.
* The State Government's inspections of safety conditions were hopelessly flawed. The Chief Inspector of Factories was dismissed from his post because he had renewed UCIL's licences every year 'without taking into cognizance', said Chief Minister Arjun Singh, 'the safety lapses in the factory'.
* None of the senior, erring State Government officials offered to resign independently or acknowledge a shred of responsibility for the accident. In this, they behaved as shoddily as Union Carbide

and its subsidiary. The only person to offer his resignation was Shyam Sunder Patidar, the Labour Minister.
* The inspectorate of factories was understaffed, its employees were underpaid. Many of its officials had neither telephones nor transport facilities. Most of them were mechanical engineers with little knowledge of chemical processes. Sometimes, they moved about in vehicles belonging to those industries they were supposed to monitor and, if necessary, penalize.
* When the inspectorate took action against UCIL's Bhopal plant in 1981 and 1982, it filed three court cases against operators and managers there. In 1987, the cases were still pending.
* The inspectorate had fifteen officials to monitor conditions in 7,000 factories statewide. According to the Central Labour Ministry, a factory inspector should not have to inspect more than forty industrial units per year.[13] In the case of Madhya Pradesh, the ratio worked out to more than 450 factories per inspector per year, clearly an impossible task.
* The Government failed in its overall response to the disaster. The army and the medical community were the only arms of the Government which appeared to function.
* On the night of the disaster, the Government could have organized mobile teams to evacuate the sick and injured to hospitals and clinics. Nothing of the sort was done until well after daybreak. But here there is a larger issue involved. Few developing countries, including India, have devised disaster-preparedness plans to deal with natural or man-made calamities. This is the case even today because national priorities have been focussed on creating employment, higher wages and achieving self-sufficiency in food and industrial production.
* The health questions raised by the tragedy have been badly mismanaged (even the sodium thiosulphate controversy remains unresolved). One obvious flaw is the lack of a single large medical centre that could cope with all the seriously afflicted.
* The Government has proved that despite its determination to extract compesation its bureaucracy in so lethargic that it mocks the Government's claim that it cares about the victims at all. Of the nearly 520,000 claims before the Claims Commissioner very few had been processed when the Government filed its case in Bhopal in September 1986. At the end of 1987, the number of

claims scrutinized had risen to 80,000 and it was calculated that it would take four years more for all the cases to be processed. Charges of corruption and fraud made the relief effort look even worse.

* The Government is directly culpable for permitting and encouraging slum dwellers to settle next to the plant.[14] The high death toll unquestionably stemmed from the large number of people living in such proximity to UCIL. This would not have happened if the city administration had implemented its own laws and ordered UCIL to relocate its plant as was suggested in 1976. Six months before the tragedy, the Town Planning Board, which was responsible for identifying danger zones in the city, categorized eighteen industries as obnoxious and suggested their monitoring. Strangely enough, UCIL was not on this list, although it was Bhopal's largest chemical plant storing huge amounts of toxic material.

This last point regarding slums around large industrial installations deserves further discussion as it is a problem that will not go away. And as long it continues to be a problem, there is no guarantee whatsoever that more Bhopals will not occur.

The problem is not restricted to India alone and is common in virtually all developing nations. And the problem is composed of so many disparate aspects that it is difficult to find a simple solution to it.

To take Bhopal as a sample case, as the city developed into the state capital in the 1960s, pressure grew on the land and encroachments began on government property. The neighbourhood of Islampur grew outward and northward toward the Union Carbide factory. It was one of the old settlements of Bhopal, a former princely state. But its renewed growth was haphazard and unregulated. This was unsurprising as the neighbourhood was increasingly composed of new migrants from the countryside who flocked to the city in search of jobs. And there were jobs available—new roads, buildings and sewer lines to be built, plus all the ancillary jobs a growing city spawns. Other attractions the city offered rural migrants was the glamour of city life and the lessening of caste discrimination which is still firmly rooted in the countryside.

Tiny hutments and ramshackle colonies sprang up along the roads leading from the Hamidia and other parts of the city to the plant. The grounds opposite the UCIL factory were listed as residential colonies but housed illegal squatters.

The Government did nothing about the encroachments. It allowed the migrants to stay, put in electricity and water lines and finally legalized the colonies.

This was because the new migrants were soon seen by local politicians as potential 'vote banks'. This was why Arjun Singh, the Madhya Pradesh Chief Minister, with an eye to the general elections expected later in the year, presented *pattas* in April 1984 to the inhabitants of Jayaprakash Nagar and other *jhuggi-jhopri* dwellers. Singh acknowledges that the motivation was political, although he defends the decision. 'It's easy to find fault with the decision and say we shouldn't have done it but anyone who knows the condition of slumdwellers in India knows they live virtually sub-human lives', he says. 'It was not meant to be a final solution (sic) of the problem but we wanted to give them the psychological assurance that (the land belonged) to them'. Singh admits too that he was unprepared for the disaster; he claims he never saw the newspaper reports filed by Keswani warning about the potential for disaster at the plant.

By giving the people of Jayaprakash Nagar the *pattas,* the Government sealed their fate. Who would reject such an offer? Land in a city and that too for free.

The victims of Bhopal never really stood a chance. For investigations have revealed that the current antagonists, UCIL and the MP State Government, had numerous links which prevented the Government from acting against UCIL. For one thing, UCIL was the showpiece of Madhya Pradesh's industrial scene. Former government officials and their relatives found gainful employment with the company. One such employee was the former inspector general of police; another, the nephew of a former minister, was a public relations officer with the company. The company made a contribution of Rs. 10,000 to a development project in Arjun Singh's constituency and the Chief Minister even used the UCIL guest house occasionally. He says he worked on office files there.

The elegant guest house was the scene of a Congress Party regional meeting in 1983. High state officials and even politicians from New Delhi stayed there during visits to Bhopal. Among these was Madhavrao Scindia, a minister in Rajiv Gandhi's Government, and a scion of Gwalior's royal family.

The link between the Government and the company was thus both covert and overt.

Occasionally, though, official agencies did act against the plant.

In September 1983, Nilay Choudhury, the National Pollution Board chief, visited Bhopal. An environmental consultant suggested that he visit the UCIL plant. He noted one major defect during the visit: the effluent evaporation pond had not been well constructed. There were reports of toxic material leaking from the pond and spreading to neighbouring land. Choudhury was concerned about the poisoning of underground drinking water sources which could affect the city. He told the factory's managers to fix the flaw. Some weeks later, he heard that the work had been satisfactorily completed.

It is easy to bring all sorts of charges against various groups in Bhopal—the Government, local environmental authorities, journalists (except Keswani), activist groups and trade unionists who weren't effective enough but that wouldn't provide the underlying reasons which led to Bhopal happening at all. We know why the Government allowed the slums around Bhopal, but it must be stressed the Government's actions were only part of the larger issues involved.

Take, for instance, the question of why there has been (and continues to be) a lack of public consciousness about industrial safety. The answer to that is that not much attention has been paid to it because it has never been seen as a national priority. Produce more, and the costs will take care of themselves, appears to be the credo the country's policy makers follow.

And what should the Government had done about the slum dwellers? Dispossessed them? No government could have done that without alienating the poor and that is a political risk few administrations are prepared to take. Besides, political considerations apart, in a country as deprived and underdeveloped as India, the Government has a responsibility to shelter the hopeless and homeless. Thus, although Arjun Singh's decision on the *pattas* may have been flawed and politically expedient, it was not morally wrong. He was seeking to answer a uniquely Third World problem in the manner of a Third World politician. Where the Government erred was in not checking the dangers the UCIL plant posed to the people living around it. A fairly simple solution to the problem of locating the slumdwellers would have been to give them government land some distance from the city. The Government could have then arranged for the installation of various facilities that would have ensured the migrants stayed on in their new surroundings instead of moving back to the city. But all this is retrospective wisdom.

What of the present? And the future?

Three years after the gas killed thousands, and injured several thousands more, life had not got better for most of the survivors. And their existence will not get any better unless the settlement being talked about is quickly finalized, properly distributed and managed. If the lethargy that marked the Government effort in the first couple of years after the tragedy persists, then the fears that any settlement might not be well-handed are very real indeed.

What is especially painful about the attitude of the Government and others down the line is that there were ways in which something could have been salvaged from the disaster. Lessons could have been learned. It is truly unfortunate that the Government, and its principal adversary, Carbide, chose to wrangle so bitterly. For in the process opportunities to succour the victims and be constructive about the future were irretrievably lost.

If India and UCC had agreed on a compensation figure in 1985, they could have launched a joint project to rescue Bhopal, assisted by major international support. Such an initiative could have transformed the lives of the victims and their city.

American management skills and resources married to Indian talents and sensitivities would have produced a rare child: a Bhopal reborn, revitalized with clean water, new sanitation and sewers, modern buildings to replace the grimy stink of Jayaprakash Nagar, playgrounds for the children, good schools and healthy food for their families and the blessing of fresh air.

There would have been conflicts of management, of cultural attitudes and perceptions. But the bitterness and rancour and hopelessness that characterize Bhopal today could have been avoided.

Because of the tardiness of the arrival at an understanding between the Government and Carbide a reborn Bhopal is still distant. A city that could have been resuscitated and become a living example of cooperation between the East and the West must instead share the fate of other cities of the developing world—busy on the surface but nursing a festering, bleeding sore underneath.

Given its hurt, Bhopal will suffer more than other cities.

# SECTION II

# A Poisoned World

In May 1984, Louis V. Planta, chairman of Ciba-Geigy, the Swiss drug and chemical multinational, denounced the enemies of transnationals.[1] He told a meeting of share-holders: 'I am referring to internal consumer protection, environmental protection and Third World organizations, and also....the World Council of Churches. Their aim is to change society. Autocratically, and often arrogantly, they lay down the law on what is good or bad for other people, and answer to no man. Their declared objectives are often merely a front for aims that are fundamentally ideological.'[2]

Companies like Ciba-Geigy are extra sensitive to criticism by the press and activist groups which question their methods of developing the trade in chemicals and pesticides. In one case, Ciba-Geigy tested Galecron, a new pesticide, on Egyptian children to discover its toxic properties.[3]

According to James Erlichman author of *Gluttons for Punishment,* 'In 1976, a group in Alexandria were paid a few pennies to stand almost naked in a field while a crop-spraying airplane loaded with the chemical passed overhead to spray them. The company then took urine samples from the children to discover how much Galecron they excreted.'[4] This might seem extraordinarily callous but it isn't the only example of what activists call 'human guinea pig' experimentation by transnationals.

Indeed, there has been, in the wake of Bhopal, a new questioning of the roles multinationals play in our lives and whether the goods they manufacture to answer the world's preoccupation with feeding itself and banishing hunger can be seen as a justification for irresponsible behaviour. That said, it is doubtful whether this new awareness is sufficient to change the way industry continues to brush aside issues of community safety and environmental protection. And it appears its only response will continue to be the kneejerk kind of reaction that Union Carbide displayed in Bhopal. This is unfortunate for there is no sign that the danger hazardous industries pose to the world at large is decreasing.

For instance, November 1986 saw a rash of leaks, spills and emissions from Ciba-Geigy, Sandoz and BASF in western Europe. These spills destroyed marine life and poisoned drinking water for miles along the Rhine. In the end West Germany was forced to provide drinking water in tanker trucks to worried citizens. The incident also soured relations between Switzerland and West Germany.

'The Bhopal of Europe' was what former West German Chancellor Willy Brandt called the first spill from a Sandoz AG chemical warehouse. The warehouse caught fire in Basel, Switzerland. Pesticides, solvents and about 200 kilograms of mercury were washed into the Rhine as firefighters battled the blaze.[5] Ten to thirty tons of chemicals may have entered the river. The West Germans accused the Swiss, and specifically Sandoz, of negligence and poor safety standards. Scientists added that a 300-kilometre stretch between Basel and Mainz in West Germany had suffered serious ecological damage. An estimated one million fish died. Some experts said the spill had destroyed years of work to restore marine life in the river, which was so heavily polluted by industrial wastes in the 1960s that more than a dozen species of fish were wiped out. Environmentalists say it could take as long as ten years for the river to recover its ecological balance.

The accident killed no humans but brought Bhopal very close to Europe, particularly the Swiss, who've always prided themselves on their efficiency and high personal and industrial safety standards. Swiss officials began looking at whether the country should change its laws on hazardous industry to conform with European Community directives on pollution control. Those directives followed a major disaster in 1976 that forced the evacuation of parts of the Italian town of Seveso after a leak from another Swiss multinational, Hoffman-La Roche.

The Basel spill has 'shattered our confidence—that of our neighbours too—in self-regulation for the industry', said the head of the information service of Switzerland's Federal Office of Environmental Protection.

'A number of far-reaching rules regulating the industry exist', he said. 'In the past they have often not been strictly enough applied by cantonal (provincial) authorities. This will change.'

The spill and subsequent leaks from other plants in Basel and West Germany resulted in scenes that have now become familiar: demonstrations and rallies demanding stricter safety in the chemical industry.

Wearing gas masks and carrying placards which read 'Today the fish, Tomorrow us', demonstrators reminded Europeans of the dangers from hazardous industries. One imaginative group even buried the Rhine with a symbolic funeral march to the accompaniment of a mournful dirge. Basel found itself dubbed Chernobasel and Bhobasel, references to the nuclear accident at Chernobyl and the tragedy in India.

Yet, unlike in the case of Union Carbide, both Sandoz and the Swiss Government responded reasonably quickly to the crisis, promising compensation and a clean-up of the damage caused by the spill.

'We have accepted full and complete moral responsibility', said a Sandoz spokesman. Another company official, Hans Winkler, told reporters that 'Sandoz and the insurance companies are ready to ensure effective treatment of demands for compensation which we receive'.

As in India, after Bhopal, Europe reported a spate of gas leaks and chemical spills after the Sandoz incident. One of the new cases involved Ciba-Geigy which leaked hundreds of litres of a weed killer named Atrazin into the Rhine, also at Basel. Yet, in keeping with the normal tight-lipped attitude multinationals have displayed in the face of disaster, the company did not release details of the spill for several days.

West German officials say they believe the chemical firm did not report the incident accurately or adequately. Environment Minister Walter Wallmann told the German Parliament at Bonn that the discharge of the weed killer was larger and more concentrated than Ciba-Geigy had first reported. Criticism of the Swiss came from another quarter, the European Community. Stanley Clinton Davis, the Community's Environment Commissioner, said that officials in Switzerland had 'accepted too readily' assertions by Ciba-Geigy that a second leak of fumes from its plant at Basel was not toxic.

In all a total of five chemical accidents were reported in November 1986 at Swiss firms. Four of them occurred at multinational installations.

If the Swiss experience by itself was not enough, West Germany discovered that half-a-ton of a herbicide had spilled into the Rhine from a BASF AG factory. The company had incorrectly reported earlier that only a quarter ton of the weed killer had been flushed into the river. The herbicide, dichlorophenoxyacetic acid, was found in concentrations up to one thousand times the level considered safe for drinking water. Two water purification plants were closed down.

The recent spate of news reports about environmental disasters caused by industry does not mean they are a new phenomenon. It only means that awareness about environmental pollution has increased to a new high. Indeed, until Bhopal occurred, the only major industrial tragedy that impinged on the world's consciousness was the disaster at Seveso in Italy. Seveso marked the beginning of a change in attitudes towards erring industry that it is to be hoped will be accelerated by Bhopal.

To get a better idea of the shift in attitudes, let's take a look at industrial disasters from around the world and assess their various aspects, how people and authorities have responded to them and what lessons can be learned from them. The Kepone leak case gained a certain notoriety in the US in the early seventies.

In April 1974, civic officials in Hopewell, Virginia, noticed a malfunction in the city's sewage treatment plant. When they investigated the incident, they realized that the plant's bacteria, which converted raw sewage to less harmful products, had been destroyed.[6] The killing of the bacteria was later traced to the discharge of a pesticide named Kepone. The company manufacturing the product, Life Science, had been granted a permit by the city administration to discharge its wastes, including some Kepone, into the municipal sewer system.

The destruction of the bacteria was noticed two months after Life Science began dumping Kepone into the sewer system.

As a result, the chemicals flowed through the treatment plant into the James river, contaminating water, aquatic life and settling in the river sediment. Plant workers were heavily contaminated and some developed severe maladies of the nervous system: trembling hands, forgetfulness, nervousness, involuntary eye movements, weight loss, sterility and pains in the joints and chest. (A number of these symptoms are similar to those still being experienced by some Bhopal gas victims.)

The Kepone plant-owners at Hopewell shut down the factory in the face of a furious official response. Significantly, and this is particularly relevant to the Bhopal case, a district judge fined the city $10,000 for failure to notify federal authorities of the Kepone discharges. The firm which first produced Kepone in Hopewell, Allied Chemical Corporation, was fined $13.24 million, a fine which was later reduced to $5 million after the company donated $8 million to the State's environment endowment.

The company which discharged Kepone into the sewer system, Life Science, reached an out-of-court settlement with fifty people, including plant workers and their families, for about $3 million. The case led to a major review of safety standards in the chemical industry by US federal and other agencies. The chemical was banned by the EPA because of worries that it could lead to cancer and also because of its permanence in the environment.

Several years later, Kepone levels were reported to have dropped but were still above national limits in several species of fish.

The chemical industry has its own view on all this, saying it is being unfairly portrayed as the villain of the piece. But the growing threat to mankind from the artificial regulation and abuse of the environment cannot let industry—chemical, nuclear, whatever—off the hook. Bhopal and Chernobyl, the two most recent examples of technology going awry underline the responsibility that industry needs to shoulder. While the scope of this book is too limited to debate the dangers of the nuclear industry, one cannot talk about the danger to the environment and communities from hazardous industries without taking a brief look at the nuclear controversy. Chernobyl saw to that.

A 'worst case' situation occurred at Chernobyl, near Kiev, in the

Soviet Union in April 1986, bringing into focus the need for nuclear safety on a global scale.[7] At least thirty-one plant workers were killed in the accident, hundreds more were injured and about 130,000 persons were evacuated and settled elsewhere; dykes were built to stop radiation from flowing into a river, old wells were sealed and new ones dug. Food in a contaminated zone across the Soviet Union and Europe was destroyed.

Experts agree that the Chernobyl disaster has prompted governments and corporations around the world to reevaluate their nuclear programmes, even though the Soviet plant is very different from others elsewhere, in technology, safety features and process systems. This is all to the good as the nuclear threat (from other sources than weapons systems) is more widespread than is commonly believed.

The *New York Times* has estimated that there are 374 operating nuclear power plants in the world. Another 157 are under construction. 'More than 700 million people live within 100 miles of a nuclear plant', the paper reported. 'About 3 billion people, three-fifths of the world's population, live within 1,000 miles, the maximum distance at which food consumption was restricted due to Chernobyl contamination.' Thus, many people live within touching distance of non-weapon-related nuclear calamity. And, as the *Times* said, in an analysis of the response to Chernobyl, 'an accident destroying a reactor somewhere in the world has a seventy per cent chance of happening in five years and an eighty-six per cent chance of happening in ten years.'

Even if one does not dispute the need for nuclear power and all the things in its favour—it does not harm the environment (if run with strict safeguards), is cost-efficient and does not deplete other natural resources such as coal and wood—one has to balance these with other aspects of the issue.

'Chernobyl, with its major operator violations, proved again that humans are the weak link in reactor safety', wrote Stuart Diamond, who studied the nuclear problem for the *Times*. Diamond, who had also investigated the Bhopal incident for the newspaper, added that the Soviet tragedy exposed 'the biggest weaknesses in accident response'. He quoted the International Atomic Energy Agency as saying that 'the general field of accident management techniques is still in its infancy'.

The world should rightly be concerned about a nuclear holocaust. It should be equally worried about relying too much on peaceful atomic power programmes as the only alternative to meet the world's energy and other needs. Nuclear power and chemicals are, without doubt, essential to man's existence. But they also need to be used responsibly. Otherwise, they could very quickly defeat the very purpose of their existence, as aids to man's survival. An early warning of this came from Seveso.

Seveso occurred eight years before Bhopal, but the two incidents bear many similarities. Both involved multinationals, manufacturing pesticides. In the case of Seveso[8], it was Hoffman-La Roche of Switzerland which owned the Givauden company, which in turn owned the ICMESA *(Industrie Chimice Meda Societa Anonyma)* firm at the Italian town. Both incidents involved the leaking of a deadly intermediate; both accidents occurred over a weekend; and, in both, plant officials and local authorities were slow to act. In fact, the plant manager at Seveso initially said after the accident that there was nothing to worry about but advised against eating local vegetables and fruit. At Bhopal, Jagannath Mukund, the works manager, told a magistrate that methyl isocyanate was 'not known to kill' and suggested that water was the best antidote to the poison.

In both cases, company leaders misled local officials, wittingly or otherwise.

Soon after noon on 10 July 1976, the safety valve on a reactor vessel containing TCP, or 2,4,5-trichlorophenol, blew at the ICMESA plant after the temperature in the vessel had soared to nearly 500°C. The result was that some 500 kilograms of TCP and about 0.5-5 kilograms of dioxin, a by-product of the chemical reaction involved in the making of TCP, spewed out of a relief pipe into the air.

Dioxin is produced in small amounts at normal working temperatures of 170-to-189°C in the TCP production process. But at higher temperatures, greater quantities are produced. Scientists describe it as one of the deadliest poisons known to the world. One gram is capable of killing several thousand people.

Seveso is not far from Milan, a distance of about thirteen miles.

The poisonous fumes affected the homes of 739 people. There was no public reaction from the company that day. It was only on 11 July that the plant's production manager made his irresponsible statement that there was nothing to worry about. On 12 July, the ICMESA plant (except for the damaged reactor vessel) resumed production. Nothing was apparently reported by the plant officials on the nature of the gas released.

'Within days, household pets in the area began to bleed at the nose and mouth and then die. Vegetation withered, almost 80,000 domestic fowl died, and so did hundreds of pigs and cattle, as well as wild birds and insects,' says an authoritative study of the incident.

The study added, 'On the afternoon of 15 July, Hoffman-La Roche rather vaguely informed local authorities that a highly toxic material might have escaped, again without mentioning (dioxin). Residents began to complain of blistering, acute diarrhoea, dizziness, headaches, and liver and kidney pains. Children suffering from burning skin rashes were admitted to local hospitals where doctors were puzzled by the malady. Workers at the ICMESA plant grew suspicious and demanded a meeting with management. When all their efforts were rebuffed, the 210 employees went on strike.'

A full week after the leak, Hoffma-La Roche announced that dioxin was involved and advised local hospitals of the threat. ICMESA's managing director and plant manager were briefly arrested for culpable negligence. Nearly another week passed before the Italian subsidiary, Givauden, after testing soil and plant samples, urged an immediate evacuation of the most contaminated areas.

Italian soldiers sped into Seveso on 25 July, cordoned off the factory, and evacuated about 700 persons. The sale and consumption of vegetables, fruit, milk, and meat produced locally were banned, and residents were advised to refrain from sexual intercourse if any of these had been consumed. As at Bhopal with Union Carbide and its subsidiary, Hoffman-La Roche's offer of major assistance to Italian officials was largely disregarded by the Government.

Activist groups called for a boycott of company products as the radicals had sought in India (but failed). Allegations that

ICMESA was involved in secret research into poison gas and defoliants for use by the North Atlantic Treaty Organization (NATO) were energetically denied by Givauden. (Similar charges against UCIL's research centre surfaced after the Bhopal disaster). But denials by the American Embassy at Rome of ICMESA involvement in American defence contracts could not be confirmed 'because order books had vanished from ICMESA's office after the explosion'.

The number of people affected by dioxin grew larger every day. A new fear emerged: that of babies being born malformed to pregnant women exposed to the leak. Large numbers of women sought special permission for abortions in a nation that is strictly Roman Catholic and bans abortions. The Justice Ministry declared the women were exempt from the laws of the land, stirring a major religious controversy. Church officials denounced the abortions.

Numerous negligence suits were filed against ICMESA on behalf of afflicted Seveso citizens. When high levels of dioxin were reported from new areas near Seveso (and 600 children with skin rashes turned up)—5,000 people brought a consolidated class action suit against the regional government authorities alleging neglect and dereliction of duty after the leak and during the decontamination.

ICMESA was not a direct subsidiary of Hoffman-La Roche. Yet, against legal advice, the Swiss firm, to its credit, agreed to settle all costs uncovered by insurance. One estimate says it paid out about $25 million in settling nearly 6,000 civil cases for compensation. The company's gesture precluded a long court battle which could have resulted in heavier compensation payments.

Union Carbide failed to show either that sagacity or sensitivity, which could have lessened the agony of Bhopal.

Seveso's effects were felt beyond the borders of the small Italian town and travelled across the country and western Europe to the Americas. The Italian Government announced tighter pollution control laws and checked chemical plants in and around Milan. More than 300 were found violating regulations. TCP units were closed or suspended operations in Britain and West Germany. A Brazilian environment protection agency banned the production, formulation and use of materials containing dioxin, Sweden declared a ban on herbicides containing dioxin and West Germany tightened controls on hazardous chemicals. A slightly defiant note was sounded by Dow Chemical, the major TCP producer in the Western world, which declared that a Seveso-type situation would be impossible at its unit in Michigan in the United States because of existing controls.

Dow's words sounded ominously similar to those of Jackson Browning, a vice-president at Union Carbide, who had proclaimed that a Bhopal-type situation was inconceivable at the MIC plant at Institute, West Virginia.[9]

As we have seen earlier, less than six months after Browning's forceful assertion, the Institute factory vented poisonous gases, sending more than 100 persons to local hospitals. This happened despite $4 million worth of new investments in computerised safety and leak detection equipment.

Thus, there may never be an industrial or chemical plant that is one hundred per cent safe. The only way they can be made safe is by vigilance, extraordinary vigilance with no place for complacency or shortcuts.

The most enduring legacy of the Seveso incident was a proclamation on industrial safety and occupational hazards at work known as the Seveso Directive[10] which was passed by the European Economic Community on 24 June 1982. The directive laid down rules member States should follow to prevent industrial accidents and to limit damage from such incidents. The sweeping legislation spelt out the steps a manufacturer of a hazardous compound must take as well as those incumbent on local authorities to ensure safety evacuation and emergency measures.

It defined the 'dangerous substances' that could be of risk to communities in western Europe and listed them under classified heads. It listed the quantity of dangerous chemicals that should be stored in industrial units.

Methyl isocyanate is number thirty-six on the list and the total amount recommended for storage is *one ton*. In Bhopal, a total of about sixty-five tons of the chemical was kept in large storage tanks and small stainless steel drums.

The directive also declared that the members of the community could extract information about hazardous substances from manufacturers relating to the exact production process, the behaviour of such substances under normal and abnormal conditions and 'the forms in which the substances may occur or into which they may be transformed'. It also said that countries should seek information which would help their local authorities deal with emergencies and prepare emergency plans for use outside the establishment.

Yet, despite such model legislation, the ghosts of Seveso as of Bhopal and elsewhere remain, for laws have yet to find a way of dealing with human greed and selfishness that obey no rules.

There is an interesting sequel to the tragedy at Seveso.

In April 1983, members of the European Parliament censured Hoffman-La Roche and the Italian authorities for allowing tons of dioxin-contaminated soil from the Seveso area to be secretly shipped out of the country, apparently headed for West Germany.[11] Proceedings of the European Parliament say that members 'were astounded at the ease with which toxic waste could be exported'.[12] A pollution control official told the parliamentarians that the commission had 'no way of demanding and even less of enforcing checks which were currently the responsibility of member States'.[13] A West German minister confessed that he did not know what had happened to the soil or where it had gone.[14]

A resolution passed by the Parliament denounced the 'cavalier attitude' of companies involved in cases such as this and urged the passage of a directive, similar to the Seveso Directive, which would control the shipment not only of hazardous wastes but also of 'all dangerous and contaminated substances'.[15]

This incident, as much as anything else, shows that good legislation and humane governments can often be brushed aside by the compulsions of multinational corporations and irresponsible governments.

The giant chemical multinationals, a bare thirty of which control more than ninety per cent of the world trade in pesticides, have prospered over the years as the demand for better methods of crop protection has grown.[16] The top ten companies, which are in the main American and West European, account for fifty per cent of the volume of the world's pesticide business.[17] Some idea of their financial power can be gathered from the fact that worldwide sales of pesticide were estimated at about $13 billion in 1983, one year before Bhopal.[18]

Of this figure, one-quarter is controlled by just three firms. One of them, Ciba-Geigy, is Swiss; a second, Bayer, is West German; the third, Monsanto Corporation, is American.[19]

Pesticides therefore is big business. The sprouting of new vectors and pests has meant that more expensive drugs must be made to fight them. Thus, growing food is getting more expensive. So is its marketing and buying. And with every hike in prices, a percentage goes to the pesticide manufacturer.

There are basically four types of pesticides: rodenticides (used against rats, field mice and other members of the rodent family): herbicides (to tackle weeds): fungicides (to combat fungii): and insecticides (against insects).[20]

The growth in demand has meant that new collaborations or licensed ventures, involving multinationals who have better expertise and funds than other companies, have soared in the nations of the developing world. This was one reason for the existence of UCIL's Bhopal plant: the Indian Government had not succeeded in its efforts to develop its own pesticides and so had opened the door to chemical companies from across the world, Union Carbide among them.

One estimate says that the worldwide pesticide industry is growing at the rate of 12.5 per cent annually.[21] And David Weir, an author who specializes in environmental issues, writes that markets have grown rapidly in the poor world. 'In Africa, for instance, pesticide use was projected to have quintupled during the past ten years', he says.[22]

Weir adds that studies have pointed to other disturbing facts: that seventy per cent of the pesticides used in the developing world are applied not to crops for domestic consumption but to those grown for export.[23] Thus, ironically, the toxins used to increase crop yields are returning to the developed world, where they originated. This also rubbishes the theory that pesticides are helping the poor feed themselves and eventually prosper. Besides, in actual fact, most of the land in the poor world, except in Communist countries such as China, is controlled by a bare three per cent of landowners. These powerful and affluent landowners, a United Nations study says, monopolize some eighty per cent of the land in eighty Third World countries.

It is relevant here to look at some of the more harmful effects of pesticides, quite apart from the dangers inherent in manufacturing them.

One of the most obvious dangers is the continuing regular and indiscriminate usage of pesticides that have proved to be harmful. This is particularly evident in the developing world, where chemicals banned in the developed world continue to be used. BHC, an acronym for benzene hexachloride, aldrin, paraquat, parathion and malathion are a few of the pesticides manufactured by multinational corporations and smaller companies in South and Southeast Asia and used indiscriminately in these and other developing nations. The pesticides are banned or used under severe restrictions in the West where most of the producers are headquartered.

Researchers who surveyed the use of chemicals in four Asian nations, Sri Lanka, Malaysia, Thailand and Indonesia, found remarkable similarities in the users' lack of training, general ignorance of the toxicity of the chemicals they were handling and ineffective government machinery to monitor the hazards and implement preventive measures.[24]

For example, farmers in Malaysia were seen to apply pesticides to their crops 'at pre-determined times, irrespective of whether or not the potential pest population (had) attained its economic injury threshold'. In Thailand, the Pesticide Poisoning Report, an activists' pamphlet, noted that peasants mixed lanate, an ingredient in methomyl, to homemade, illegal brew 'in the belief that it (proved) more "kick" to the drink.'

The Western companies which sell hazardous pesticides to the Third World aren't dubious, fly-by-night operators. Among the twenty-eight firms which market pesticides restricted in the West, but not in the developing world, are eighteen American companies, five West German firms, two Japanese corporations, two British and one Swiss company. There are names such as Bayer, Chevron, American Cyanamid, Dow, Hoechst, Montrose, Shell, ICI and Union Carbide.

Their Third World cousins are no less guilty. Paraquat, the weed killer which can cause suffocation when inhaled or splashed on the skin, is manufactured by no less than ten Taiwanese firms of a total of seventeen top producers worldwide.[25] The other producers include a Mexican and a Malaysian company.[26] D.D.T. is manufactured by one major Indian company, the government-owned Hindustan Insecticides, which has a factory in a densely populated area of New Delhi, and by companies in Brazil, Argentina and Colombia.[27] HCH is manufactured by eight Indian companies in both the public and the private sector and firms in Mexico, Brazil and Argentina.[28] HCH is a suspected carcinogen and has never been registered for use in the United States, according to the Pesticide Network International report.

Even where governments in the developing world know of the dangers of banned pesticides, they hesitate to do away with them for they are the cheapest and most effective means to enhance crop production which is seen as a key to economic development. Thus in Indonesia—as in other developing countries—the Government heavily subsidizes the use of agrochemicals, creating bigger markets for large companies like Monsanto and ICI whose shining factories continue to spring up in the developing world.

And as with governments everywhere, it isn't only the public weal or the lack of money to develop environmental safeguards that makes the authorities encourage the growth of multinationals in the world. There are the usual stories of kickbacks and other forms of shared partnerships between local governments and big industry. As a British activist remarks, about the relationship between the chemical industry and government in his country, 'They are in and out of bed with each other the entire time. You see this at conferences. Whitehall officials and industrialists are obviously comfortable in each other's company, not least because they have spent a lot of time serving on the same committees. The ministry will always bend over backwards in the long run to give industry the benefit of the doubt where safety is concerned.'[29]

This last remark underlines the fact that the danger the chemical industry poses is not to the developing world alone.

The National Resources Defence Council, a leading American environmental forum, says that a 1984 study reported high residues of pesticides in nearly half of California-grown fresh fruit.[30] The residue turning up in the fruit was the ubiquitous D.D.T., banned in the United States in 1972. According to Drew Douglas, an environmental writer, an upsurge in D.D.T. levels has been reported from parts of California, New Mexico and Arizona, after the pesticide's levels had dropped nationwide earlier. 'And most troubling of all, officials in (the) States—along with federal investigators—are unsure how or why the contamination levels (are) rising', writes Douglas.[31]

It is undeniable that multinationals manufacturing pesticides and allied material do have a role in development. They do and must continue to play a part as was evident from the success of the Green Revolution that, in the Indian subcontinent alone, spread prosperity and higher productivity over wide areas.

The danger governments and farmers must realize is in their attributing higher crop yields and returns to pesticides and fertilizers alone. Good crop management, better irrigation and power facilities, the sensible use of land, the proper choice of crop, the right selection of strain—all these are essential for agricultural success stories. But pesticides now have a growing allure, especially in the fast-growing markets of Africa, Asia and Latin America, in no small measure due to large scale advertising by pesticide companies, and are regarded as the only answer to bigger and better crops. Apart from the dangers to man, the fact that heavy use of pesticides only makes pests more resistant and even immune (through mutation) is often forgotten. This is in the manufacturers interest, of course, for the demand for a more potent product never ends.

More than a quarter century ago, an ecological prophet had predicted the shape of things to come. In her compelling masterpiece, *Silent Spring,* Rachel Carson spoke of the futility of unleashing chemical warfare on insects and life systems that have evolved over millions of years and of the need to 'share our earth with other creatures'.[32]

'As crude a weapon as the cave man's club, the chemical barrage has been hurled against the fabric of life, a fabric on the one hand delicate and destructible, on the other miraculously tough and resilient, and capable of striking back in unexpected ways. These extraordinary capacities of life have been ignored by the practitioners of chemical control who have brought to their task no high-minded orientation, no humility before the vast forces with which they tamper', wrote Carson.[33]

Twenty-five years on, the practitioners of chemical control haven't changed their basic attitudes.

Pesticides are only one aspect of the vast chemical industry that intrudes into the daily lives of virtually every one of us. Chemicals are used in cosmetics, food and drink, drugs, industrial products and fabrics.

In the United States alone, more than 12,000 chemical companies have annual sales of more than $182 billion and employ more than a million people.[34] According to one account, of an estimated five million chemicals that exist on earth, there are nearly 66,000 in use in the United States.[35] These include 48,523 in commerce; 3,350 pesticides registered with the EPA; 1,851 prescriptions and over-the-counter-drugs and excipients approved by the Food and Drug Administration (FDA); 8,627 approved food additives; and 3,410 chemical ingredients used in cosmetics.[36]

Carl Pope, a leader of the conservation and environment protection movement in the United States, has estimated that a bare eight per cent of all food additives, ten per cent of all pesticides and one per cent of all commercial chemicals have been carefully tested in his country.[37] Apparently, the proliferation of chemicals 'overwhelms both testing capacity and regulatory process'.[38] He says that between two and four years and anything between $40,000 dollars and $1 million is required to test each chemical.[39] Given all this it is obvious that a lot more needs to be done before we can safely co-exist with chemicals. There are signs though that an effort is being made.

Multinationals have shown they are capable of a voluntary and effective role in the fields of environmental protection and industrial safety, Bhopal, Seveso and other industrial disasters notwithstanding. According to one report, big business is funding nearly forty per cent of the total spending of the OECD countries (Organization for Economic Cooperation and Development) on environment protection.

After the United Nations Conference on the Environment at

Stockholm in 1972, multinational corporations as well as governments have developed environmental policies which assess product hazards and effects on the natural environment.

This is not entirely a new approach because European producers were behind the development of safeguards in the aluminium industry as early as 1900. The aluminium industry was particularly concerned about the exposure of workers to fluoride, a by-product in the manufacturing process. By the 1940s the industry was reducing fluoride emissions and integrating controls into the production process.[40]

Other instances exist where industry has made major contributions to environment protection. For example, Philips, the Dutch electronics manufacturer, which began as a maker of light bulbs, has set up a special plant in the Netherlands to test and manufacture products that are safe for developing nations. To take another example, the 3 M Company coined the phrase 'Pollution Prevention Pays'. At a conference on waste reduction in 1984, 3 M reported on its pollution prevention programmes in twenty countries. 'These programmes have resulted in the elimination of over 140,000 tons of sludge, and over 1 billion tons of waste water', the report said.[41] As a result of this, the company saved an energy equivalent of some 254,000 barrels of oil per year. And the most telling part of the whole exercise was that these results were achieved not through the use of pollution control equipment but through better, cleaner, manufacturing technology.

Yet, despite extolling the virtues of pollution prevention, 3 M turned down a request by business colleges in Manila which wanted to study the clean technology in the company's plant in the Philippines.[42] The decision raised a question whether the technology used by the parent American firm had reached its Filipino unit.

Pope, the US environmental activist, makes the point that while companies agree that products should be safe and pure, there are few industries which would withhold commercially viable chemicals from the marketplace 'simply because testing data is incomplete, or because there is some evidence that the product may be toxic in some circumstances!'[43] This isn't the same as the public view which naturally does not want to be at the mercy of harmful products. As Pope says: 'People want chemical products that will not cause injury when used as directed'.[44] They do not want to be told years later that the product can cause birth defects or harm pregnant women. But, Pope adds, 'the chemical industry does not accept the legitimacy of the public's desire for safer chemicals and larger margins of error'.[45] It resents government regulations and curbs and resists public fears. This it does through 'administrative appeals, political pressure, and lawsuits which slow down the regulatory process; and the Government's energy and resources are exhausted long before the problem can begin to be solved.[46]

This is true of the chemical industry anywhere in the world.

As we've seen before, no one can seriously dispute the role of transnationals in the economic growth of the developing world. Equally, no one can deny the fact that quite often MNCs have one standard governing their behaviour in the West and another in the developing world. Their plants in developing countries are usually not as efficient as similar plants in the West, nor do they follow laws as carefully. Interestingly, their pollution records are often better than those of local firms. There are few studies comparing the industrial pollution records of multinationals and local firms but those that have been done provide interesting insights. Let's examine three studies in Malaysia and the Philippines, in the early 1980s.

In the Malaysian study, conducted by Gregory Thong Tin Sin, about 100 local and multinational companies were studied on the basis of interviews with senior managers.[47] For the physical environment section of the questionnaire, most foreign companies scored above average (the scoring was done by Sin with 100 per cent as the top score): for example, Singaporean companies scored eighty-one per cent; Japanese, eighty-eight per cent; and British, American and Australian, seventy-five per cent. Malaysian firms scored below average, with seventy-one per cent, and Dutch companies had only twenty-five per cent.

The study assessed British companies and local ones and reported that the foreign firms did better in air and noise pollution research than in water pollution and energy conservation. The British spent more on air pollution control. The multinationals were seen as concerned about legal responsibilities but on average not behaving very differently from local companies. The report concluded that, 'overall, MNCs are no worse—and are marginally better than local companies in terms of environmental awareness-...Given the evident technological and financial strengths of the multinationals, compared with the local companies, one might have expected an even better performance by MNCs, and hence one could perhaps conclude that, in this instance, the performance of the MNCs is disappointing.'

In the Philippines, thirty companies, including fifteen multinationals, were studied in fourteen industrial sectors. The inquiry was conducted by Juanita Manalo, the head of the Science Department of the Philippine Women's University. It covered a ten-year period, between 1972 and 1981, and the companies were given a possible score of one or zero on four points: whether they were polluting when the study began; whether local residents had complained against them; whether they had been fined for violation of national pollution control laws; and whether they were still polluting at the end of the study.

In the scoring, multinationals had fewer bad points than the local companies (nineteen to thirty-three). All the local companies were reported to be causing pollution at the start of the study period, compared to two-thirds (ten) of the MNCs. In terms of complaints by local residents, these occurred more than twice as much with local companies as with the MNCs. No MNCs were fined for violations, whereas two local firms were. Finally, MNCs did better in solving their pollution problems. At the end of the period, only five MNCs had not resolved their problems, whereas nine of the local companies still had difficulties. This trend is perhaps more the rule than the exception. Yet a Bhopal, a Seveso or other instances of industrial catastrophe blows the lid off the reputation of high environmental safety standards multinational companies claim to set for themselves.

One last example will suffice to show the extent of danger from irresponsible industry.

Minamata is a name as closely associated with tragedy as Hiroshima or Nagasaki in Japan. In the 1950s, residents of Minamata, a fishing town on the Japanese island of Kyushu, had begun to notice strange changes in the bird and marine life of the bay where they fished.[48]

Fish kills were reported as far back as 1950. Two years later, crows and sea birds began dying. By 1953, worried inhabitants saw their cats going mad and dying after suffering a grotesque dancing disease. Cats would stagger around as though drunk, salivating; then they would suddenly be shaken by convulsions and whirl about involuntarily before collapsing and dying. Some threw themselves into the sea or rivers, others into kitchen stoves. By 1957-58, no cats were to be found in four parts of the Minamata area. Despite the fish kills and the inexplicable behaviour of the birds and animals, fishermen still went out to catch fish in Minamata Bay and nearby areas and people continued to eat the fish.

In April 1956, a six-year-old girl entered the factory hospital of the Chisso Corporation in Minamata with symptoms of brain damage: difficulties in walking, disturbed speech patterns and delirium. A few weeks later, her sister and four neighbours were found suffering from similar ailments.

Chisso, a producer of nitrogen fertilizers, had been discharging methyl mercury, a by-product of an industrial process involving the use of inorganic mercury for manufacturing acetaldehyde and vinyl chloride, into the bay for years. The company had first built the factory at Minamata in 1907 and had begun producing acetaldehyde in 1932.

In May 1956, the head of the Chisso hospital declared that 'an unspecified disease of the central nervous system has broken out'. Investigations by health and factory authorities located 30 such cases and noted that most of the patients had been sick since 1953 and lived in the fishing villages of Minamata. Their ailments had been diagnozed incorrectly until then, with physicians saying they were suffering from syphilis, alcoholism and encephalitis.

In mid-1956, a special research group was set up to investigate the strange affliction. Within two months the researchers made public an interim finding: that the health problems were not being caused by an infection but by heavy-metal poisoning caused by eating infected fish in Minamata Bay.

Yet neither the local Government nor Chisso acted responsibly. Despite the committee's discovery, that heavy-metal poisoning caused by contaminated fish was responsible for the maladies, the local administration did not ban fishing in the bay. Nor did Chisso stop its production until the cause of the problem was pinpointed.

Two years after the medical investigation had begun, a Japanese medical specialist noted the similarities between the poisonings at Minamata and those reported in 1940 at an English factory producing methyl-mercury. Tests on cats who were fed with methyl-mercury showed the same symptoms as those which had danced to death in the Japanese villages. The following year, scientists who took samples of mud from near Chisso's drainage channel, which flowed into Minamata Bay, found exceptionally high levels of mercury contamination. Autopsies on those who had died turned up high concentrations of the metal. Victims who were still alive showed high levels of mercury in their hair.

The mercury deposits discovered in the autopsies were high enough to cause severe kidney, liver and brain damage because organic methyl mercury passes through the blood-brain barrier more easily than inorganic mercury.

Chisso then announced that it could not possibly be the culprit since it used only inorganic mercury. It began to oppose the investigations and refused to give sludge and factory waste to the researchers.

By a stroke of good fortune, a scientist found a bottle of sludge from the acetaldehyde process lying in a laboratory. A study of the waste showed the presence of methyl-mercury chloride. There was now no question that it was indeed Chisso that had made the stuff which had poisoned the sea, the fish and those who ate the fish. Yet, after a while, people began to eat the fish again because they felt action was being taken to keep the 'disease' under control. Incredibly, the Government did nothing to stop them or move against Chisso.

Officials and villagers in the area then began reporting another alarming development: women who had eaten infected fish during their pregnancy bore children who were congenital idiots, with severe mental and physical retardation. They were unable to coordinate limb movements, had deformed limbs, sometimes lost consciousness and were otherwise disoriented.

Yet, Chisso did not stop its discharge of methyl-mercury wastes until 1968 when the Government officially announced that the corporation was to blame for the disaster. It was later revealed that when the initial outcry had begun, Chisso had diverted its waste products into the Minamata river briefly in September 1958, thus contaminating parts of the Shiranui Sea beyond the bay.

An analysis of the overall failure to press ahead with the identification of more disease victims between 1960 and 1971 points to five factors:

(1) Inhabitants of the Minamata area simply did not know how they were to apply for official verification of their health problems.
(2) The stigma of the malady was such that no one wanted to be identified as a victim.
(3) The fishermen feared that their fish would not sell and kept quiet.
(4) Many thought their ailments were from other causes and that the Minamata malady was a thing of the past.
(5) The Government failed to pursue any research worth the name.

In fact, the official machinery failed to move in the matter—as at Bhopal—until the efforts of individual activists revealed that the effects of mercury on the Minamata population were far wider than earlier believed. By the end of 1974, officials had identified nearly 800 victims, of whom 107 had died. There were nearly 3,000 others who were applying for verification of their claims.

While all this was going on, twenty-nine families of victims in Minamata filed a consolidated compensation suit against Chisso. This followed a division among the victims. One group had opted for a settlement under which the Government was to settle the case on their behalf and fix compensation amounts. The other group, charging the Government with equal culpability in the affair, refused to go along with the settlement and went to court.

After four years, the district court at Kumamoto awarded the victims compensation. Chisso announced it would pay $3.2 million immediately. The court divided the victims into three categories: those with severe cases of injury received about $10,000 each; the less severely damaged were to get about $9,500 and those with other afflictions would receive about $9,200.[49]

However, after-verdict consultations for a mutually acceptable settlement continued for nearly three months. By 1975, Chisso had paid out indemnities totalling more than $80 million.[50] It was also paying all medical expenses, plus monthly allowances of $60-$180 per patient depending on the degree of illness.[51] This is a good yardstick to calculate compensation for the Bhopal victims.

Raj Kumar Keswani, the journalist who had predicted the Bhopal disaster, travelled to Minamata in December 1985. He reported then that new victims of the malady were still turning up, that some of their cases had not been decided and they were preparing to demonstrate before the country's Environment Ministry.[52]

Keswani says he has failed to draw a dividing line between Minamata and Bhopal. 'It's all the same', he says.[53]

Minamata forced the Japanese Government to form a Central Pollution Board. Victims of mercury poisoning were present at the UN Conference on the Environment at Stockholm. Their moving testimony and presence helped shape the call for the protection of nature's ecological systems and to prevent hazardous dumping of wastes, a call that has led to scores of nations developing environmental law and companies fashioning pollution-control policies and devices.

But it can never be stressed enough that a lot more needs to be done. And the chemical companies have enough profits at their disposal to set up state-of-the-art R&D and pollution control facilities that could ensure that further damage is kept to the minimum. This, of course, is a meaningless point to make unless everyone concerned—industry, government, consumers—decides that it is best to work toward an environment that is as harmless as possible to the people who live in it, even at the cost of immediate gain. Until that thinking begins to gain root, nothing very constructive can happen.

# India's Neglected Environment

On the fourth of December 1985, just one year after the Bhopal disaster, Mahesh Mehta, a lawyer specializing in cases relating to environmental pollution, was preparing to appear before the Supreme Court in New Delhi and argue for stricter controls on industrial pollution in Gujarat and Maharashtra. As he waited for his case to come up, Mehta strolling outside the splendid, high-domed building noticed a large number of lawyers running.[1]

'We are going to die', one of the lawyers told Mehta. Their words were an echo of those spoken by the residents of Jayaprakash Nagar just after the toxic gas enveloped their settlement at Bhopal.

Perplexed and worried—the memory of the brutal anti-Sikh killings in 1984 by mobs avenging the assassination of Prime Minister Indira Gandhi by Sikh guards was still fresh in the public mind—he asked the frightened men what they were running from. The response was that there had been a gas leak in West Delhi but the lawyers were unclear about the details.

Mehta, a small, bearded man who lapses from Hindi into English when he gets excited, went to the offices of the *Indian Express* a short distance away to track down the cause of the panic. He found the place virtually deserted but ran into Coomi Kapoor, an enterprising woman reporter, who said the gas had leaked from a factory, in crowded North Delhi, belonging to the Shriram family, one of India's biggest industrial familes. But, again, she had no details.

Mehta's next move was similar to that of Keswani in Bhopal. He telephoned the police control room which assured him that the situation was under control.

He returned to the Supreme Court where he told the judges about the leak. 'They jumped', he recalls and asked him to tell them what he knew about the incident.

Elsewhere in the capital, there was total panic as residents, fearing a second Bhopal, abandoned their vehicles and fled from the pungent fumes, identified later as a mix of oleum and other gases. Massive traffic jams added to the fear. Cars collided and markets closed. More than 300 people were hospitalized, of a total of about 700 who were affected by the gas. But only one death, of a Sikh lawyer the following day, was directly attributed to the leak from the Shriram Food and Fertilisers plant.

All India Radio put out announcements saying the leak was not toxic. (This was incorrect, scientists say). And police vehicles went around the city proclaiming there was no danger. Three top plant officials, including the manager of the factory, were arrested and charged with negligence. An official inquiry committee was set up to probe the incident and fix responsibility. The Supreme Court also appointed four committees in succession (the first three made recommendations which were either contested by Mehta or the company) to study the issue. A fifth was set up separately by the local administration. Members of Parliament attacked the administration for laxity and Shriram for low safety standards.

The leak occurred when badly corroded struts holding up a tank of oleum, a highly concentrated mixture of sulphur dioxide and sulphur trioxide, gave way and the tank crashed to the ground.[2] The force of the impact severed a connecting pipe and about forty-two tons of oleum flowed out. The initial reaction of labourers working in the area, almost overpowered by the powerful "stench" from the tank, was to toss buckets of water at the chemical gushing out.

It was the worst thing they could have done.

Water has an exothermic or heat-producing reaction on oleum and liberates sulphur dioxide, sulphur trioxide, sulphuric acid and oleum. Sulphur dioxide is recognized as a major health hazard. The Central Pollution Control Board in India has set the following levels for its presence in the air: in industrial areas, 120 microgrammes per cubic inch; for residential areas, 80 microgrammes and in sensitive or protected areas, 30 microgrammes. More worrying, sulphur dioxide converts readily into the trioxide, which is even more poisonous.

The procedure which should have been followed was quite different: the contaminated area should have been isolated by circling it with sand and earth; the spill should have been removed by a vacuum truck. The area should have then been washed with water, followed up by further neutralization of the water with lime and soda ash.

When the workers threw water on the spill, they noticed the acid was changing into thick white vapour. Breathing became more difficult, and they turned to flee out of the factory gate. But the gate was shut. So they ran to another area of the plant which was unaffected by the wind.

The situation was worsened by the arrival of the fire brigade, which turned its hoses on full blast at the acid spill, sending an enormous cloud of acid gases into the air and across the crowded neighbourhoods near the plant. Coughing and wheezing, the people of the area began to run, seeking escape.

They were lucky that it was not chlorine which leaked but oleum. For chlorine was stored at the plant as well, in six tanks. It was used for producing caustic-chlorine-soda and *vanaspati*, or dehydrated vegetable cooking fat.

In fact, local officials had been pressing for the closure of the company's chlorine plant which they felt posed a threat to neighbouring residential communities. Two vocal members of Parliament had urged its immediate closure saying a leak from the chlorine unit would cause a tragedy worse than Bhopal because about half-a-million people lived in a five-kilometre radius around the Shriram plant. Chlorine can incapacitate and kill if inhaled in large amounts.

The Delhi Administration shut down the plant after the 4 December leak.

On 6 December, a second but smaller leak of sulphur dioxide was swept by a strong breeze to the city's downtown area and large numbers of people felt its sting, although no injuries were reported. About half-a-ton of sulphur dioxide had escaped when a safety valve failed as oleum was being transferred from one tank to another. The leak was controlled but not before a fresh scare shook the neighbourhoods around the Shriram plant. The Shriram case became a landmark in environmental litigation in India.

Interestingly, until the Shriram incident occurred, a full year after the Bhopal disaster, and alerted bureaucrats and politicians in New Delhi to the possibility of their being gassed in the capital, the Government had not come forward with a law to prevent industrial accidents of this nature. But after Shriram, the Supreme Court in India introduced the concept of managerial liability in the event of an industrial mishap and marked the judgement as a model for regulating hazardous plants nationwide.

The leaders of Shriram, which had a turnover of Rs. 5.18 billion in 1983-84[3], and is one of the country's better-regarded industrial groups, at first sought to minimize their responsibility as Union Carbide had done. Yet, several months later, to their credit and in contrast to Union Carbide's stubbornness, they accepted the interim Supreme Court ruling.

The change in attitude did not occur overnight.. Initially the company contested several parts of the crucial ruling. But bowing to judicial, executive and public pressure, it agreed to sign the conditions of liability which the court had sought.

It was a remarkable moment in the country's legal and corporate history and followed a three-month tussle in the courtroom. It came two years after the first major warning of the dangers the plant posed had been sounded by a British expert. In that report, the expert declared that the managers were not operating the plant safely and responsibly and said it could be classed as a major hazardous facility. He added that the Government should consider constraining the plant's activities to protect the public and employees and concluded that only relocation could guarantee a risk-free operation.

After the accident, all four committees appointed by the court criticized the functioning of the factory. While a committee picked by Mehta (at the court's suggestion) urged the relocation of the plant out of Delhi with a minimum ten-kilometre green belt—or uninhabited zone—around it, it also attacked the factory's undersized neutralization systems and unsafe working conditions. These charges were stoutly denied by the company.

Siddharth Shriram, the company's deputy managing director, made one important point before the investigations began.

'Assuming we shift our factory, where is the guarantee that we will not be asked to move once again'? he asked.[4] The new location, he argued, could turn into a residential area as well. He was underscoring a point which was first raised by the Bhopal disaster—of population pressures that result in crowded slums and housing colonies coming up near industrial plants.

Shriram pointed out that the area where the plant was located had once been a jungle. The dwellings came after the factories were built. Mehta in turn said that under the Delhi Zoning Act the area was supposed to have only medium-sized industries. Shriram was violating that law by its very presence since it was a giant enterprise sprawling over seventy acres, he said.

The company said that the closure of the factory was costing the company heavily and added that the subsequent production loss of *vanaspati* was also affecting the market and causing hardship to consumers. Company officials also pointed out that the city's water supply undertaking depended on Shriram for its chlorine. They added that if Shriram was to be closed permanently and moved out, then the same standards should be applied to Hindustan Insecticides, the government firm which manufactures D.D.T.[5]

Nilay Choudhury, the former chief of the Central Pollution Control Board, said he had once feared a major accident at Hindustan Insecticides, which he described as shabbily maintained with leaking pipes and joints.[6] The factory was also a user of Shriram facilities: the chlorine from Shriram was piped one kilometre to the government firm. But after Bhopal, and the accident at its neighbour, Hindustan Insecticides was shut down for several months as experts studied potential risk areas in the plant and suggested ways to remedy the problems. After Bhopal, a survey of high-risk factories had identified both Shriram and Hindustan Insecticides as users of dangerous substances.

After studying the reports of the committees, and listening to the company's arguments, the Supreme Court on 17 February 1986 pronounced an interim order: Shriram could reopen its caustic soda and *vanaspati* plants but not the oleum unit.[7]

The court laid down eleven conditions for the operation of the plant.

The key conditions of the judgement related to personal,

managerial liability for accidents, a concept untouched by Indian tort or industrial law. It said that managers of the company would be held culpable in the event of future accidents. They were to deposit Rs. 1.5 million each as bank guarantees.

In other parts of its ruling, the court denounced the ineptitude of the Delhi Municipal Corporation for failing to clean the sewage drains in the area for five years, forcing Shriram to empty its toxic wastes into the Najafgarh drain, which supplies irrigation water to farmers in the area.

The court ordered the appointment of trained safety officers in each of the factory's units and recommended surprise weekly inspections of the plant by factory inspectors or trained chemical engineers and the installation of notices and loudspeakers to warn the public of future leaks.

It ordered Shriram to deposit Rs. 2 million as security towards the settlement of compensation claims and urged the Government to set up a body which would supervise the functioning of hazardous industries. It also said that it had faced great difficulty in procuring the services of independent experts who could study the questions the accident raised. The judges said the Government should institute an independent Ecological Science Research Group which could act as an information bank for the court and government departments. The Government should consider setting up environmental courts as well, comprising a professional judge and two experts from the Research Group. Appeals would go directly to the Supreme Court.

In March 1986, the court responded to complaints by the company managers who said that fears of liability would keep responsible people from joining the company by modifying one of its conditions to say that only the senior officer running the caustic chlorine unit would be responsible for any future compensation claims arising out of a leak of chlorine. He would have to pay his entire annual salary to the victims.

However, it also declared that it had difficulty accepting a company argument that the managing director and chairman should not be held responsible for accidents on the plea that they were not involved in the daily running of the plant. 'We do not see any reason why the chairman and/or managing director should not be required to give an undertaking to be personally liable for payment of compensation...particularly when we find...there is considerable negligence in looking after its safety requirement.' Only such a clause, the judges said, would ensure proper and adequate maintenance of safety systems.

The company responded courageously: Siddharth Shriram accepted unlimited liability even though such a condition was not even mentioned in the ruling. This assurance was accepted by the court although Shriram awaited a final ruling upon appealing the judgement. The Shriram case showed too that when pressured Indian companies could act responsibly.

This was not the final word on the Shriram case. In December 1986, Chief Justice Bhagwati ended his last day in office with a flourish, by establishing a new concept of managerial liability—'absolute and non-delegable'—for disasters arising from the storage of or use of hazardous materials from their factories.[8] It was a benchmark judgement for hitherto, as Bhopal showed there was no such ruling on industrial liability in India.

'We are of the view,' the ruling read, 'that an enterprise which is engaged in a hazardous or inherently dangerous industry, which poses a potential threat to the health and safety of the persons working in the factory and residing in the surrounding areas, owes an absolute and non-delegable duty to the community to ensure that no harm results to anyone. The enterprise must be absolutely liable to compensate for such harm and it should be no answer for the enterprise to say that...the harm occurred without any negligence on its part'.

The court's ruling said that hazardous processes were necessary for development, although the court wanted the least risk of danger to the community.

Justice Bhagwati said later that it was not a new principle that the court was emphasizing. He was responding to the question of whether Carbide could take the plea (in the Bhopal case) that it was being unfairly subjected to retrospective law-making. He also said: 'Whenever any case of tort arises or comes before the court, the court has to lay down the correct principle on which liability has to be determined and compensation evaluated'.

He and his fellow judges believed that hazardous processes could endanger life and therefore infringe upon the 'right to life', the most basic of rights. That was why, Justice Bhagwati said, company management would be totally culpable for such accidents or mishaps—for no individual corporation or government had the authority to take the life of any person.

He said later that the judgement applied equally to government-run firms that cause pollution or damage to life and property, not merely to private sector companies like Shriram and Union Carbide.

The response from industry was not favourable. One chemical industrialist said he was scrapping plans to open three new chemical firms and considering divesting from the three he already controlled. 'I don't want to end up behind bars for any accident that takes place and which is not within my control', he said.[9]

An industry spokesman was even more scathing. 'The judgement means that the court wants the chemical industry in India to close shop', he said. 'There is nothing like a one hundred per cent safe plant, everyone knows that—we have to live with risks, with reasonable risks.'[10]

Pressure will grow in the coming years from the chemical industry—private and public sector—to force the Government to dilute the effects and beliefs behind the Shriram judgement. It will be an indicator of the administration's commitment to the safety of its people and industry's acknowledgement of its moral responsibilities if the judgement continues to be in force.

The judgement and the company response clarified one thing: that there is little point in trying to shift existing polluting or hazardous plants out of a city. The threat of relocation was a useful weapon with which to warn an offender but it would be impractical to implement. 'What we need to ensure instead is that new plants which endanger communities or the environment will not be allowed to come up without inbuilt, adequate, extensive safety and pollution control measures which are constantly up-dated', says Dilip Biswas, a director at the Ministry of Environment. 'The old plants must be made as safe as possible by adopting new technologies, even though this may be expensive'.[11]

When we look at the question of industrial safety in India, there are several issues that have to be considered. India has now achieved self-sufficiency in food production, and has outgrown its begging bowl image of the 1960s (though the drought conditions in 1987 blurred the picture). Yet, nearly half of its people live below the poverty line and exist on one proper meal a day. The cities represent to rural populations, especially the younger generation of labourers and farmers, an escape from poverty and unemployment. Mass migrations take place. Urban slums come up (the slums around UCIL in Bhopal, for example, were predominantly composed of migrant labourers).

To feed, house and give its people work is the responsibility of the nation-state. One way of achieving this is by increasing the industrial capacity of the country and initiating new projects which can generate employment. In a nation like India, where politics is usually slanted towards appeasement, the Government cannot risk the disaffection of the unemployed. Nor can it afford to be casual in its treatment of the vast professional and industrial work force—engineers and lawyers to mill workers and unskilled labour. Their disgruntlement can easily be turned against the State by ideologies of the left or right. So the factories will continue to come up. Indeed, India, and other developing countries, which have vast labour forces and limited job opportunities, face a bitter dilemma where environmental issues are concerned. Simply put, this dilemma is: Can poor countries with limited resources and creaking infrastructures spend heavily on controlling hazardous industries when these very industries give employment, albeit at a certain cost, to the community. How much of a risk is riskable? At what level does it become unacceptable?

'To judge by the international response to the Bhopal tragedy...one would think that the accident occurred in the US, rather than in India', commented *Business India,* a Bombay-based magazine in mid-1986.[12] It summed up the Government's response as modest: it had filed a suit for damages against Union Carbide, made vague noises about the need for comprehensive regulatory controls in hazardous industries, and enacted an Environment (Protection) Act. 'What has precisely been achieved, is, as yet, unclear', said *Business India.*[13]

It compared the Indian effort to the spate of proposed legislation in the United States after Bhopal: eleven bills had been tabled in the US Congress seeking to amend legislation relating to environmental and occupational hazards; four Congressional committees had held hearings on the competence of laws regulating toxic chemical manufacture, transport and emissions into the atmosphere, occupational safety and health as well as export controls on American products and technologies.[14]

Compared to that, India's effort certainly seemed meagre.

To go back to the beginning India's first step in officially combating pollution and protecting the environment came soon after the 1972 Stockholm Conference on the Environment. That very year, the Central Government set up the National Committee for Environmental Planning and Coordination. This was followed by the creating of pollution control boards at the Centre and in the states. By 1980, most of the states had their own pollution control boards and that year the Central Government also formed a separate Department of Environment. In less than seven years, the department saw five Secretaries or top bureaucrats come and go. This high turnover of officials was scarcely likely to create confidence among the staff or help settle important issues relating to the department's work. That said, the setting up of the Department of Environment and the promulgating of the new Environment Protection Act were welcome developments. The new Act vested the responsibility of improving the environment and controlling pollution in the Central Government, taking away the existing authority of state administrations. It also empowered the Centre to set up a Central Environment Authority which would act as a nodal point for all environment-related government activity. It put the Government's decisions on these issues beyond the jurisdiction of civil courts and gave the authorities the power to force the closure of any industry, operation or process found polluting the environment. The penalties for offenders also became harsher: a maximum prison term of five years for a first offence and a fine of Rs 100,000. An additional daily penalty of Rs 5,000 could be levied for as long as the violation of rules continued after the conviction. An offence which continued a year after the conviction could result in a seven-year prison term.

The power to shut down an offending plant was a bold and long overdue step. The law also conceded the rights of the citizen to battle polluters. It allowed any person to move the court against a polluting company, provided he gave it sixty days notice and informed the government agency concerned (normally, the pollution control board) about the offence. A court could not admit the motion, however, if an official agency had already moved against the firm or told the court of its intentions.

The Government sought to protect itself against suits saying no legal complaints could be made against it, or officials and authorities appointed under the Act, who had done anything 'in good faith.' But that was a small matter to concede for until the law was passed, there were no laws which enabled Indian citizens to file complaints against polluting companies and agencies; they had to rely on public interest litigation, a concept championed by former Chief Justice P.N. Bhagwati of the Supreme Court.

However, there were those who saw certain provisions in the law hampering litigation by radicals and concerned citizens. 'The ban on challenging officials in a civil court is aimed at reducing the role of Indians who are worried about pollution', says Mehta, the lawyer who fought Shriram. He also criticized the sixty-day limit, saying it virtually pre-empted any serious complaint against a company because the Government could quickly take over the case.

Environmental specialists say the rest of the law is merely a rehash of existing legislation. It does not specify procedures or set pollution limits. It only enables the Government to frame measures to deal with these problems, although both the Air (Prevention and Control of Pollution) Act and the Water (Prevention and Control of Pollution) Act are more detailed on polluting emissions.

The Environment Act provides for the setting up of five major environmental laboratories to test and monitor air and water samples. Until these actually begin functioning, the Government must rely on sample analysis conducted at private or government laboratories handling other scientific work.

It cannot be denied that the Act is, unfortunately, not comprehensive. For instance, there still is no list of hazardous substances used in the country. A move to introduce such a list was dropped by the Environment Ministry in the summer of 1986 after the Government decided to push the Environment Act. According to environmental officials, politicians and bureaucrats felt the new law would take care of concerns about dangerous substances because it enabled the Government to legislate on these issues.

The new law on pollution control is one of over two score legal weapons that the Indian State has assembled to deal with environmental pollution, industrial safety and workers' health. Some of these laws are more than a hundred years old, such as the Mines Act of 1852. Others concerned with hazardous processes include the Indian Explosives Act (1884), the Factories Act (1948), the White Phosphorus Act (1933), the Hazardous Occupation (lead) Rule (1937), the Organic Solvent (rubber) Rule (1937), the Industries Act (1951), the Prevention of Food Adulteration Act (1954), the Insecticides Act (1965), and the Prevention of Water and Air Pollution Acts (1974). Of nearly fifty laws, about twenty-two deal specifically with environmental pollution.

The most widely used have been the Air and Water Acts. But although prosecutions by the state pollution control boards have been numerous, there have been few actual convictions. The process is frustrating and long because there are no special environmental courts yet (despite the Supreme Court's suggestion) handling the cases. They must be heard with the raft of cases already existing in overcrowded civil courts. Adjournments are frequent and largely initiated by the prosecuted company, whether government-run or privately owned.

Choudhury, the environmental scientist, says that he knows of virtually no major convictions under the Water and Air Pollution Control Act. The only one he cites is of a court in Rajasthan which, in 1983, nine years after the Water Pollution Act became law, awarded a two-year prison term and a Rs. 2,000 fine to the manager of a paper mill. 'The prison term was bailable but even if the manager goes to jail, the proprietor remains a very honourable man, he is not touched', Choudhury says. The new Environment Act sought to close that loophole by saying that neglect on the part of a director or secretary of the firm would make such officers liable for prosecution.

The real test of such ambitious legislation will be its implementation by government agencies, acceptance by industry and interpretation by courts. An assessment of how well it works will not be really known for another decade. By then, it is to be hoped, courts will have developed and refined environmental law, and the Government, pollution boards, trade unions, industry and environmental groups clarified their responses.

A review of earlier environmental problems, particularly in the chemical industry, indicates the magnitude of the problem in India and how much needs to be done to resolve it.

India's chemical industry is about 100 years old. It began with the setting up of oil refineries. Inorganic fertilizers were the next major step in the growth of the industry followed by synthetic dyes, drugs and pharmaceuticals, pesticides, petrochemicals, thermoplastics and man-made fibres.

According to a 1986 Indian Government estimate, the chemical industry produced goods that year worth Rs. 200 billion or about $17 billion.[15] This was ten per cent of the Gross National Product that year. Of this, fertilizers accounted for Rs. 80 billion, oil and fats Rs. 50 billion, and pesticides and bulk chemicals another Rs. 50 billion. Drugs, petrochemicals and polymers accounted for the rest. The industry's share in India's gross national industrial output in 1983-84 was placed at forty per cent, a five-fold jump over the eight per cent it contributed in 1970-71. Yet, despite these impressive figures, the average Indian uses and buys only one kilogram of plastics every year, compared to the per capita consumption of 40 to 50 kgs in the West.

Safety standards, however, have not kept pace with the accelerated growth of the chemical industry. In the late 1980s the industry has 4,000 big factories, with a large number in thickly populated areas: it also has the dubious distinction of being the most dangerous industry in the country. In 1980 alone, 100 workers were killed and 10,000 others injured in accidents. In 1979, India reported 37.33 accidents per 1,000 workers in the chemical industry.[16] In the United States, the comparable figure was 14.2 accidents pér 1,000 workers and in Britain it was placed at 0.50 per 1,000.[17]

Of the 4,000 chemical units, 326 are classified as manufacturing petrochemicals, pesticides, inorganic and organic chemicals, dyes, dye intermediates and drugs. The 326 plants are among the chief high-risk industries in India.

But according to one of the country's best known environmentalists, they are only the tip of the iceberg. Anil Agarwal of the Centre for Science and Environment, a pioneering group that has produced two masterly documents on the state of the national environment, says that there are tens of thousands of small, illegal factories in towns and cities which violate existing laws with impunity and use and manufacture dangerous chemicals. Many of these are not covered by the Factories Act, which is applicable only to units employing more than ten persons.

While the chemical industry is dangerous, it is also crucial. The Indian petrochemical industry supplies much of the country's fertilizers and pesticides, which help the country feed itself. The country's pesticide production has soared from 154 tons in the 1950s to a current level estimated at about 78,000 tons. According to Agarwal, about seventy per cent of this amount is contributed by pesticides banned or restricted in the West, including D.D.T. and BHC.

The Government is evaluating the need for these two pesticides, which have played a central role in the nation's successful agriculture story, as well as for other pesticides which have been branded harmful, toxic and the use of which is highly restricted elsewhere.

One spin-off of the rapid growth of chemical industries has been the proliferation of factories handling hazardous products in heavily populated cities like Bombay. According to the Garg Committee, headed by Dr. R.K. Garg, an eminent atomic scientist, there are as many as 6,000 polluting units in Maharashtra.[18] Half of them are located in and around the Bombay region. 'This figure does not include the thousands that operate illegally', says one magazine commenting on the findings of the Garg Committee.[19] The committee was set up immediately after the Bhopal disaster to assess the possibility of such accidents occurring in Maharashtra.

Maharashtra was among five states that responded to a Central

Government directive that all the 22 states and seven union territories (that formed the Indian Union at the time) study and report possible Bhopal-type situations in existing hazardous units. After two years had passed, only five states—Maharashtra, Madhya Pradesh, Gujarat, West Bengal, and Tamil Nadu—had responded. This goes to show how little political, industrial and bureaucratic concern exists for industrial safety and planned development.

This attitude needs to change if Indian industry is not to be responsible for several more deaths and disabilities. While industry in the country has certainly progressed a great deal since India became independent over forty years ago, the working class still remains underprivileged. Their living conditions are often poor, except in the housing colonies built by thoughtful employers such as the Tatas in Jamshedpur. Working conditions are little better. Thus, the incidence of ailments like silicosis, byssinosis, asbestosis and pneumoconiosis is very high.

According to a mining specialist, more than one million Indian miners suffer from silicosis, which is caused by dust containing free silica or silicon dioxide. In Mandsaur, Madhya Pradesh, an estimated 3,500 persons have died of the disease in a period of twenty-five years, a result of their working in state units.

S.R. Kamat of the King Edward Hospital in Bombay, who has succeeded in improving the health of several Bhopal victims, says that a five-year study of three textile mills in Bombay showed that twenty-four to twenty-five per cent of the workers in old or semi-modern mills suffered from byssinosis.[20] This is caused by the inhalation of cotton dust in the blowing and carding rooms. The textile industry is a major employer in India, providing about one million jobs. Many of the country's old textile mills, and some of its modern ones, are situated around Bombay. Byssinosis is said to be chronic among textile workers who have worked in the mills for more than ten years.

In this industry, as in others, accurate data is difficult to obtain. However, independent surveys in Bombay, Delhi and Ahmedabad have reported levels of byssinosis among textile workers as high as seven to eight per cent. A higher incidence has been reported in other studies which have included Kanpur, Madras, Madurai and Nagpur in their sweep.

Asbestosis, which is caused by exposure to the silica dust in the mineral, is a major hazard among asbestos workers and others in related industry. According to Kamat, more than one-third of all workers in the asbestos industry suffer from one form or another of asbestosis, which can lead to cancer. He and others say that private companies are among the worst offenders in this respect. The National Institute of Occupational Health said in a study that 224 workers out of a total of 800 in a Bombay firm making asbestos-reinforced cement showed an advanced stage of asbestosis. The management denied the report.

Barry Castleman, an American environmentalist who has specialized in exposing how American multinationals are taking the asbestos industry abroad because of pressure at home, reported in 1981 that workers at the Shri Dijvijay factory near Ahmedabad cleaned or picked up asbestos fibre with their bare hands and removed the sludge from effluent canals without wearing protective clothing. The waste was often dumped outside the factory walls and poor hutment dwellers used it for their homes. Significantly, in 1986, Manville Corp., the Ahmedabad company's American collaborator, settled a multi-million dollar compensation suit brought against it by Americans in the US.

In an attempt to improve its safety record in India, Dijvijay spent $1.7 million on a water pollution control project and other occupational health requirements.[21] Manville also sent a technical team to suggest improvements. One of its recommendations was that asbestos scrap piles on the road leading to the factory should be buried, not because they posed a health risk but because they were 'unsightly and could be photographed and used detrimentally against all asbestos cement producers'.[22]

P.K. Rane, the certifying surgeon of Maharashtra's Medical Factory Inspectorate, says it is difficult to implement safety controls and modernize plants. Partial correction of dust control is expensive, he says, and old plants have little space to realign or modify their machinery. Further, trade unions and even government agencies oppose the modernization of plants if an update of technology involves retrenchment of workers.

He says that even workers do not insist on proper monitoring at these plants but seek cash benefits and do not cooperate with rational safe practices. Dr. Rane adds that local engineers lack the expertise to modify installed equipment and make pollution-control possible.

These are the on-site conditions for workers in some of India's biggest mills and industries. Thus, although Bhopal drew national and international attention to the hazards obtaining in a modern plant, there are older industrial problems which cannot be overlooked.

One positive aspect of the Bhopal disaster and the Shriram leak was that the Indian media became much more environment conscious. Accidents resulting from exposure to hazardous substances and processes in the chemical industry were reported meticulously and frequently. A look at these reports shows that industrial accidents in India occur with alarming frequency.

Between mid-December 1984 and June 1986, at least sixty persons were reported killed and more than 530 injured, in about twenty-four accidents involving leaks from tanker trucks, factories, home industries, fireworks explosions, mine blasts and those caused by faulty cooking gas cylinders. The worst incident was reported from Karnataka's Bidar District in November 1985, when forty villagers were killed after a petrol tanker truck exploded in an accident. There was even another leak at Bhopal in June 1987, when escaping ammonia gas from a factory near the UCIL plant, caused a panic but no casualties.

Union Carbide referred to one of these incidents, a gas leak in Bombay that injured more than 150 persons including a Member of Parliament, in its reply to the Indian Government's complaint in the New York court. In this instance, about three tons of chlorine escaped from a rusted storage tank at the Illac company, a subsidiary of the Sarabhai group based at Ahmedabad (the conglomerate makes sarees, fabrics, chemicals, plastics, paints and dyes). At that time, Datta Samant, the most powerful trade union leader in the Bombay region and a Member of Parliament, was addressing a meeting of workers near the factory gates. Samant and 150 others were hospitalized after inhaling the toxic gas.

The incident occurred in August 1985, five months after the

Garg Committee had surveyed fifteen major factories in the Bombay region and pronounced virtually every factory as lacking adequate safety measures. One of these was Illac. Another was UCIL's chemical plant at Chembur (interestingly, this plant was up for sale but had difficulty finding buyers). A third was a giant fertilizer plant owned and operated by the Central Government, Rashtriya Chemicals and Fertilizers Limited. These plants and the others studied were located at Chembur, also nicknamed 'Gas Chamber' by its residents. The air in the area, located on the national highway connecting the business capital of India to Pune and the South, is heavily polluted by emissions from industrial units.

Illac was warned by the Garg Committee in April 1985 about its unsafe conditions. 'The highly-corroded valves, pipelines and storage vessels in the caustic soda and chlorine plant need to be replaced immediately', said the Garg report. It pointed to six other major flaws in the system and declared that it was not safe enough to be operated. A month later, the plant was shut down. This was not because the management was concerned by the dangers posed by the plant but because they had not paid electricity bills worth Rs. 10 million to the Tata Electric Company. The Tatas simply cut off the power. Most of the workers were laid off and company officials claimed they had written to the State Labour Commissioner of the need to remove the chlorine. Whatever the truth of these claims, nothing was done about the chlorine, and two months later it leaked.

After the leak, the company decided to neutralize the remaining stocks of the gas by injecting liquid caustic soda into the storage containers. According to Nithin Belle, a reporter with *Bombay* magazine the Government had woken to the danger in a typical manner: 'Bolting the stable door after the horse has escaped'.

The lack of official urgency in the Illac case was reminiscent of the casual way that senior bureaucrats in the labour department at Bhopal handled a report relating to the first major accident at the UCIL plant. They did not touch it for months and then read it a month before the 1984 catastrophe without taking any action.

But it is not just private industry which is at fault. Large and medium-sized government plants have also been major offenders.

According to an Environment Ministry official, what is distressing is that many public sector units are worse in their pollution management and safety features than private factories.

The leak at Illac was preceded by at least two incidents at the government-owned RCF, one of India's pioneer fertilizer plants (it was set up in 1959), involving the release of ammonia, sulphuric acid fumes and sulphur dioxide. Until the spring of that year, RCF had been described by environmentalists and residents of Chembur as the main polluter in the area. The leaks in March and April 1985 rammed the criticism home.

The first ammonia leak prompted the Maharashtra Pollution Control Board, headed by C.D. Oomachen, a state legislator, to order RCF's closure until it had fixed the fault. A month later, when the plant had resumed operations, a crack in the sulphuric acid unit sent gas fumes into the air. Workers fled and slum dwellers rushed out of their huts near the plant. Ironically, at that precise moment, the chairman of RCF was giving a talk on industrial pollution where he blamed 'poor work ethics, infrastructural deficiencies, low maintenance standards and unreliable testing systems' for the poor standards of pollution control.

The Garg Committee submitted its report to the Government later that month and devoted a good part of its attention to Rashtriya Chemicals, noting it had sales of Rs. 4 billion in 1983 and profits of Rs. 500 million. It had studied the plant twice, once in December 1984, and again in February 1985, to review the company's implementation of interim measures suggested by the experts.[23]

It listed seven flaws after the first visit: some operators were unfamiliar with safety devices and procedures for handling abnormal operational problems.[24] Ammonia, carbon monoxide and nitrogen dioxide were not being continuously monitored; safety valves and pressure indicators were corroded; there was no stand-by procedure to meet an emergency in the methyl amine plant; the ammonia storage tank needed proper checking and painting.[25] It noted that the loading and unloading of sulphur was done manually; the old nitric plant needed a new pollutant control system which would halve the level of nitric oxide emissions from 2,000 ppm to 1,000 ppm.[26] RCF's managing director, Duleep Singh, agreed to the conditions, given the public outcry at the time over pollution. The Garg committee spoke approvingly, in its final report, of the progress taking place at the plant. Despite these assurances, the leaks did take place later in 1985. According to Nilay Choudhury, the nitric acid technology was about thirty years old and needed updating. In 1986, RCF found Spanish technology to reduce its emissions of nitric acid, which is dangerous when absorbed or inhaled.

Officials at the Directorate-General of Factories Advice Service and Labour Institute in Chembur, say that RCF improved its pollution record only after it was confronted with a threat of closure by state agencies. They say RCF then spent Rs. 250 million on improving its emission levels.[27] The directorate, better known as DG-FASLI, is the chief training and research institute for industrial labour in India. It is funded by the Central Government.

The DG-FASLI is one of the best-informed institutions on labour practices and welfare measures in India. In a 1985 report, it said that fatalities and injuries in industrial accidents had been gradually increasing. Since 1980, the total number of persons injured was listed at 316,532. Of these 657 proved fatal.[28] The number of accidents in 1983 was 349,254; 864 died in these incidents. The provisional figures for 1984, excluding Bhopal, were 302,726 injuries and 824 deaths.

The greatest number of accidents and fatalities were reported from Maharashtra, the state with the highest level of industrialization. In 1980, the overall figure for injuries was 86,829, inclusive of 135 fatalities. The accident rate soared to nearly 100,000 the following year and finally came down to 67,495 with 103 deaths in 1984. West Bengal, run by a Marxist Government since 1977, has reported industrial accidents which have hurt more than 70,000 persons every year since 1980. From 73,774 that year, state labour officials pegged the number of injuries in 1984 at 81,758 including 95 fatalities.

Maharashtra and West Bengal were followed by Gujarat, Uttar Pradesh, Madhya Pradesh, Tamil Nadu and Karnataka. These states reported eighty-seven per cent of the total industrial injuries nationwide.

The directorate also spoke in its report of the burden on the understaffed departments which are supposed to inspect factories in the states. 'In almost all the states...the number of inspectors is inadequate necessitating more than 150 factories to be inspected by an inspector', it said. In some states, the directorate complained, many posts in the inspectorates were not filled because of the poor pay and facilities offered.

That was perhaps an understatement, as the Bhopal tragedy underscores. According to officials in the directorate, factories are supposed to be inspected twice every year to check compliance with safety rules. In 1984, only Orissa, Chandigarh, Tamil Nadu and Dadra Nagar Haveli said they had inspected more than ninety per cent of the factories in their jurisdiction. The rate for other states ranged between eighty and ninety per cent for Maharashtra, Gujarat and Kerala and below twenty per cent for Bihar and Delhi. Madhya Pradesh reported an inspection rate of forty to sixty per cent of all factories in the year of Bhopal.

The problems seem insoluble given existing conditions. In 1984, the number of factories that a single inspector had to cover ranged from 1,011 in Bihar to 800 in Andhra Pradesh, 508 in Karnataka, 297 in Tamil Nadu, 288 in Madhya Pradesh, 208 in Maharashtra and 327 in Delhi. Clearly these are impossible tasks.

'If you give an inspector 800 factories to survey in a year and if he needs to visit some of the bad ones several times to check compliance with rules, if you give him few facilities, no transport, no telephones, and very few instruments to carry out scientific tests, if you tell him he himself must prosecute the cases that he registers, then what results can you expect'? asked one official at the DG-FASLI.

The official adds that budgets and the numbers of inspectors have gone up in the states since Bhopal occurred, especially in places like Maharashtra and Gujarat. Also up are the number of inspections. In Madhya Pradesh, the number of posts for factory inspectors was increased from twenty-six to forty-six but the problem of getting good staff continues.

There is no doubt, though, that the low priority given to safety in industry has changed to a degree after Bhopal and the Shriram leak.

The Indian Government held meetings of labour ministers and secretaries from all the states to look at existing legislation governing safety, especially the Factories Act. It asked the International Labour Organization to draw up model legislation to govern occupational safety and health. It appointed expert groups to identify hazardous units in every state and picked a high-level committee to study forty-eight such units divided equally between the public and private sectors. The Committee was instructed to suggest ways to minimize risks to workers and communities.

These studies are important in formulating policy even though they were completed two years after the Bhopal disaster. However, their implementation is the real test, say officials and environmentalists. The Government also finalized its legal response to the problems caused by dangerous substances and processes in the form of major amendments to the Factory Act, which now has a new chapter on occupational health and covers toxic substances, their storage, manufacture and transport.

In an effort to interest private industry in investing in pollution control, the Government announced tax concessions, depreciation and investment allowances on control devices and systems. Further, it is exempting capital gains from tax in the cases of those industries which while relocating have to sell property or machinery. The Government now allows a seventy per cent rebate on the water cess levied on industries using water for manufacturing processes if local authorities are satisfied with pollution control facilities.

A Central Government announcement after Bhopal listed twenty types of polluting industries which would need environmental clearance for siting. Until then, none of the industries, each one of which is a potential killer, had required any such permission or official scrutiny of their location. The industries listed included pesticides, cement, asbestos, refineries and foundries. Such industries must now seek approval from three environmental agencies (the Department of Environment, a site selection committee and the local pollution control board) before beginning to build.

The Central Government also set up a series of committees to look at the problems posed by plants handling hazardous materials. One group set up to identify the number of units handling hazardous industries reported a total of 996 such plants. They ranged from pesticide formulators to oil refineries, space research centres and pharmaceutical manufacturers.

Most of them were located well inside the municipal limits of major cities and townships. But the list was, in fact, an incomplete one. It referred only to giant-sized and medium-sized factories. There was no mention in this report of the hundreds of tiny units in the small scale sector, many of which are not even registered, but are capable of wreaking much havoc.

The need for a list of industries using dangerous substances had also been felt much earlier.

In the 1970s, the Central Government had listed twenty-six dangerous manufacturing processes. These included the manufacture or manipulation of dangerous pesticides, electrolytic plating, manufacturing and treatment of lead and asbestos.

Some of the recommendations of the various committees set up by the Government are worth mentioning here if only because the points they mention must be taken seriously if there are to be no more Bhopals. One of the committees, headed by D.V. Kapur, the then secretary of the Ministry of Industry, surveyed conditions at forty-eight plants in the chemical, petrochemicals and drug industries and made a series of suggestions in September 1986 to improve internal plant conditions.[29] It also laid down parameters for siting.

The Kapur Committee made twelve major recommendations. Among others it said that chemical plants storing hazardous material above specified quantities must have plans to deal with emergencies inside and outside the factories. It suggested safety audits on plants handling hazardous chemicals by an independent team every three years and frequent counterchecks by official agencies. It said that regulatory agencies must rank old and new installations on a 'hazard potential' test. For new plants handling hazardous substances, a 'safety audit by an external agency was recommended prior to the plant start-up'.

A competent in-house safety unit was another recommendation.

This would include a well-staffed technical services unit and medical team equipped to monitor the long-term effects of hazardous substances on workers. All plants handling acute toxic materials should possess or secure data on the long-term effects of those substances. The expert team said that it was concerned to find that in most of the plants the technical services department was the weakest link in the chain of safety management.

It suggested the identification of potential disaster areas nationwide and the creation of a master plan for each site. It proposed that a twenty-two-point checklist of information should be available on each site including the size of population in the neighbourhood, history of conflicts with local groups on health and safety issues and whether the population at risk was aware of the dangers of living near the plant.

In its main recommendations, the committees referred in some detail to the problems caused by populations around toxic industries. 'While effective communication methods have to be established with populations around the existing companies, it is equally important to restrict further development of colonies in the vicinity of hazardous installations'. It also suggested that populations around the plant must be informed of the risks from the factories, and of what to do in times of danger. Company managers should involve the local administrations in such communication, the committee suggested.

Coming to the core of the problem, the committee outlined three alternatives for existing hazardous plants. One, reducing the amount of dangerous chemicals in their inventories. Two, shifting parts of the manufacturing process that could cause accidents and hazards to safer locations. Three, shifting the entire plant to a new site.

The last choice, the committee said, was relevant only to small and middle-sized plants in residential areas. Relocation was impractical for the larger factories. On top of the expense of transporting the plant and employees, the company would be heavily penalized by excise inspectors for moving every major piece of equipment. So, the Kapur panel said, it was better that the Government provided industries with tax incentives and subsidies. Those which needed to relocate should also be given handsome incentives.

The committee stressed the fact that new hazardous installations should be ringed with a safety buffer zone or green belt of trees and shrubs; no people should be allowed to settle within a kilometre of an industrial estate housing hazardous plants and state governments must form an industrial buffer for the protection of citizens: a band of non-hazardous industries and warehouses posing a physical barrier between hazardous units and residential colonies.

The committee said that it had listed about 1,500 industrial units, which employed more than 1,000 workers each, which needed skilled safety personnel. Current training programmes for factory inspectors and safety staff were inadequate and it asked the DG-FASLI to improve and increase manpower levels. Funding and facilities needed beefing up. Chemical engineers needed to be inducted into the inspectorate, it said, and data on accidents, safety and hazardous materials computerized.

It cited transportation of bulk chemicals as another major hazard. 'No effort is made by any company to prepare a blueprint for transport emergency and no assessment is also made on the most suitable mode of transport of a particular chemical based on the minimum damage concept', it added. It did not refer to Maharashtra's pioneering work in this field (which we shall look at later) and suggested that the manufacturer be made responsible for safe disposal of spills and leaks occurring during transport.

On small-scale industries, which constitute a major hazard in city neighbourhoods and bazaars, the committee suggested that all such factories be brought under the Factories Act, whatever their level of employment. One qualified inspector should be appointed for every 200 small units and inspectors must be adequately equipped with monitoring devices.

Another important point it made was the proliferation of laws on occupational hazards in different industries. It suggested a reappraisal of existing regulations and agencies and urged a coordinated approach which would eliminate duplication, encourage pooling of expertise.

There was need for a national board on industrial safety and hazards which would be the final arbiter in deciding policy for all process industries, including chemicals, petrochemicals, fertilizers, cement and paper. It listed twelve functions for such a board including safety audits, formulation of guidelines on chemical safety and safety prerequisites for granting licences to chemical-related industries. Similar state-level agencies were also recommended.

The siting of industry also was the focus of a special study, commissioned by the Environment Ministry, which said in August 1985 that no industry, whether hazardous or not, should be located less than half-a-kilometre from a highway or railway line.[30] This is a suggestion that has rarely drawn any interest from industry. The major industrial estates in New Delhi, Okhla Industrial Estate, and New Okhla Industrial Estate, begin a few hundred feet, if that, from large highways. One of them is only a few feet from a major railway crossing and station.

Another major recommendation made related to industries and habitation. After defining a major settlement as one comprising about 300,000 people, the report said that industry should be located at least twenty-five kilometres from the projected growth boundary of the town, ten years from the time of its siting.[31]

The illustration of Faridabad, once projected by Nehru as an example of the new industrial culture of India, tells of the pitfalls in the implementation of such a concept. In the 1950s and 1960s, when Faridabad was growing, much of the area between it and Delhi was jungle, scrubland or farms. Today, the 22-kilometre stretch between Faridabad and Delhi is lined with industries and homes on both sides of the road, in some cases barely a few feet off the national highway that goes to Agra and beyond. It is a pattern that is seen throughout India and its urban sprawls, in the Bombay-Thane-Pune belt, in the Calcutta-Howrah area and in Madras. So mere laws and compliance with them by industry isn't the answer to this problem. The real problem is the flood of migrants from the rural areas to the cities. Unless this is checked and job opportunities created in the countryside and smaller towns, the cities will always remain centres of industrial hazards.

What of industry's response to the myriad laws? Unsurprisingly it has not been very good and where industry has complied it has done so grudgingly.

Take, for instance, a test case in the development of the

relocation idea in Maharashtra, the key chemical producing state in India. More specifically, in Bombay. The Maharashtra Pollution Control Board under Oomachen ordered seven major industrial units to shift out of the Chembur area because they were considered hazardous. Two of the units, Illac and UCIL, closed down operations. The others continued to operate but appealed the order. They included three large Government factories: Rashtriya Chemicals and Fertilizers, Hindustan Petroleum Corporation Ltd. and Bharat Petroleum Corporation Ltd. According to UCIL's Vijay Gokhale, the petrochemical plants were not only dangerous in themselves but had spawned greater risks by building colonies for their workers in the area. The interesting thing is that UCIL, under a cloud since Bhopal, made a fuss about closing down. 'The Government of Maharashtra told us to move out', says Gokhale.[32] 'We told them, "Move these people (the colonies around the plant) out—they came after we did, create a buffer zone of 200 metres and we will give you a safer plant".[33]' He says the Government has told his company to move from Chembur before any accident occurs but that the cost of relocation is unmanageable: about Rs. 550 million.[34]

Despite the various laws on paper, few states have actually bothered to move in a concerted fashion against the industry. Maharashtra is one of the states which has and the chief of the state pollution control board, C.M. Oomachen, takes great pride in pointing this out. Another praiseworthy aspect of the pollution control effort in Maharashtra is that the state's commitment to environment safety started long before Bhopal. It was the first state to enact laws on water and air pollution (1974), set up a pollution control board and more recently promulgate a law covering the transport of hazardous materials (1984).

The last law places the responsibility of safe transportation of dangerous chemicals on the company producing or handling such substances. It also says that no vehicle should transport hazardous goods unless it observes several rules. Under the Act, all vehicles carrying hazardous goods must display labels representing the nature of the substance and its potential hazard; instructions on emergency response to leaks and spills must be carried by drivers (the instructions must be in Hindi, English and Marathi and the languages of the state of transit and destination); the driver has to carry a TREMCARD (Transport Emergency Card) summarizing these instructions.

The law on transporting toxic substances became necessary because the pace of technology had outgrown existing laws, which dealt with such substances as petrol, explosives and cooking gas cylinders. The rules did not touch a vast array of toxic and corrosive substances (some of which are neither inflammable nor explosive) which were as deadly as those listed upto then.

The Maharashtra law was enacted more than ten years after the United Nations Committee on the Transport of Dangerous Goods evolved a classification of hazardous substances and the International Maritime Organization developed guidelines for the storage of hazardous substances in ships.

Given the fact that seventy per cent of all road accidents in India involve transport vehicles, the law was particularly timely. In fact, according to the National Transportation Planning and Research centre, India has only one per cent of the world's vehicles but accounts for about six per cent of all global road accidents. There are an average of 200,000 accidents every year and the current accident rate of thirty-five per 10,000 vehicles is more than three times the rate in Western countries. R.T. Atre, the president of the Indian Roads Congress, says that more than 30,000 people are killed in road accidents annually.

Oomachen admits that despite the various laws enacted and the measures taken to implement them by his Government there is a lot of ground still to be covered.[35] He says that a significant measure in the right direction was the Central Government ruling that twenty-one industries, including chemical and thermal plants, should get environmental clearances from state administrations before they could begin their projects.

The Maharashtra Pollution Control Board is sometimes held up as an example to other errant boards. Nilay Choudhury recalls that he sent a team in 1983 to a government fertilizer plant in Bihar after receiving complaints from workers there of heavy pollution and medical problems. His investigation found that the ailments were caused by fluoride contamination. Choudhury says his team made thirteen recommendations to improve plant operations.

'The management implemented the two easiest', he recollects. But Choudhury does not blame the managers. 'I think the state pollution board is responsible for not following up the report', he says.

Reports from around the country suggest that even in states like Maharashtra the rate of progress in implementing environmental safety measures is painfully slow. This is not especially surprising given the magnitude of the task to be tackled. However, there are some initiatives by the community that have been encouraging. The Antop Hill incident was one.

'This project will make Bombay a Bhopal', proclaimed the September 1986 cover story of *Business India,* a business magazine out of Bombay.[36] The report spoke of a massive project to store hazardous chemicals in the Antop Hill area of Bombay, surrounded by some of the city's most congested localities. There is a population of 1.5 million living within a one-kilometre radius of the proposed facility.

The project was sponsored by private business, the Bombay Municipal Corporation and the Maharashtra Government, thus making industry and politicians partners in an ultra-hazardous facility.

The Antop Hill Warehousing Company Limited was conceived in 1975, according to Ramu Pandit, the secretary-general of the Indian Merchants Chamber, as a solution to problems caused by the haphazard and often unauthorized storage of dangerous chemicals inside Bombay. As with all such projects, the area was virtually uninhabited in the 1970s. It was considered a far safer zone than the traditional chemical market in downtown Bombay where scores have died in chemical blazes. Approving the plan, the Government decided to lease nineteen acres of land in Antop Hill exclusively for the storage of hazardous substances.

The Indian Merchants Chamber promoted the complex, which was to cost Rs. 340 million, and comprise sixteen large warehouses. Ten of these would store extra hazardous chemicals. A large number of safety features were to be incorporated in the warehouses, including the storage of large amounts of water, eight-inch-thick reinforced concrete walls capable of holding a blaze for four hours, large courtyards, verandahs to enable mobility of personnel and access to storage units in the case of fire, automatic water sprinklers. The movement of chemicals was to be handled by trolleys and lifts; there was to be a fire station, an electric station and a telephone exchange.

Arun Subramanium, who reported the dangers of the project, wrote that the authorities and private entrepreneurs had apparently not learned the basic lesson of Bhopal: that hazardous chemicals should not be stored in large quantities. (This, in fact, is one of the main recommendations of the Garg Committee which underlined the importance of restricting the storage of 'intermediate products which are hazardous in nature...to eight hour requirements, and before any plant is shut down all such intermediate products should be consumed'.)

The AHWC proposed to store thousands of tons of chemicals in drums and cylinders. The magazine said the idea was unacceptable, as such procedures could only be used for temporary storage and to facilitate transport. The article went on to talk of the dangers of contamination and runaway reactions, which could prove uncontrollable, given the large volume of chemicals proposed to be stored. The storehouse, when completed, would have the capacity to hold 50,000 tons of toxic substances.

The project was faulted for presuming that fire was the main hazard, ignoring the increased risks posed by violent reactions that different chemicals could spark off when stored near each other. The AHWC classification of the proposed goods as hazardous and extra-hazardous was described as inadequate and meaningless for it violated the guidelines that the International Maritime Organization had set for the transport of gases, inflammable liquids and solids and toxic and oxidizing substances in ships. (These standards are regarded as the best guidelines available on storing toxic material.)

The magazine wasn't being irresponsibly sensational in its disclosures. In the past, the Bombay Port Trust had seen three of its godowns destroyed in fires because of the improper storage of highly reactive substances. The lessons of these blazes apparently struck home and the Port Trust is building a new storage facility where it is incorporating the guidelines of the Maritime Organization.

In the event of a disaster at Antop Hill, an emergency evacuation of the entire neighbourhood population would have been an impossible task, made doubly so by the fact that the main routes out of eastern Bombay pass very close to the site.

In the event, the outcry generated by the *Business India* article forced the private and government promoters to reconsider the project. On 5 November 1986, the Indian Merchants Chamber announced it was stopping work on the site. But it pointed out that there was a more fundamental problem to be resolved.

'This is part of a larger issue', said Pandit. 'You have to manufacture hazardous chemicals, there's no getting away from that. But suppose we relocate even thirty miles out of Bombay, where is the guarantee that the Government will be able to keep the reserved zone free of squatters'?[37]

Echoing the frustration that Siddarth Shriram had shown earlier, Pandit said that without the political will to oust migratory settlement, new industries would always be swamped by new colonies, which would live in constant danger.

'Every politician wants to jump into the fray and say he is on the side of the deprived, but what about development, what about the commitments to us—that is the real tragedy,' he said.[38]

The Antop Hill project showed that despite their usual indifference to such issues, government and industry could be provoked into corrective action. But only after a sharp prod. It also showed up the warts on regulatory agencies, such as the Department of Environment, which had allowed the project to progress unhindered for years.

It is an indication of how much more needs to be done.

Again, as in the previous chapter, one cannot talk about the dangers of hazardous industries without speaking of the nuclear industry in India. While a detailed analysis of the Indian nuclear industry would merit a separate book to itself, it is helpful to look at some of its aspects.

The country has three major power plants located at Tarapur, near Bombay, Kota in Rajasthan and Kalpakkam in Tamil Nadu.[39] Under an ambitious project drawn up by the Department of Atomic Energy, India plans to set up a network of 235-

megawatt and 500-megawatt nuclear power reactors which will produce 10,000 megawatts of power by the end of the century. At current levels, nuclear energy accounts for about five per cent of the total power generated in India. The plan is to increase this proportion to ten per cent by the year 2000.

Even today, India's nuclear programme is extensive. It includes research facilities, plants, and a nuclear fuel fabrication centre reprocessing spent fuel into high-grade plutonium. This plutonium can either be used for a weapons programme, or in fast-breeder reactors. Despite the country's expertise there are questions about its ability to contain the inherent hazards of nuclear plants. Scientists have admitted to major contamination at Tarapur, the pioneer plant which was built in the 1960s by General Electric and Bechtel Corporation. Workers at the plant were exposed to radioactivity levels several times higher than internationally accepted levels in the late 1970s, news reports said.

There was even a possibility of a reactor core meltdown in 1980, according to *Sunday* magazine, which could have triggered a catastrophe like Chernobyl.

The danger was barely averted. Although officials now say that radioactivity levels at Tarapur have been greatly reduced and are now within tolerable limits, environmentalists believe it is still one of the world's most contaminated nuclear centres. Flaws in operation, design and equipment have caused frequent shut-downs, cut power production and lowered the plant's efficiency. Tarapur has rarely worked at more than fifty per cent of its installed capacity (it has two reactors of 210 MW each).

It has some positive aspects to it as well. It generates power more regularly than the other atomic power stations commissioned later. It produces electricity at rates lower than either the plant at Rajasthan or the one in Tamil Nadu. 'But this jewel in the Indian nuclear crown is also the world's worst performing nuclear station', says the Centre for Science and Environment, citing the previously-listed complaints.

The other plants have their problems too. The facility at Kota has functioned only spasmodically, with long interruptions caused by construction and materials problems. The first reactor there has been shut for two years and may never be reactivated because of a major crack in the structure which could endanger the project and communities around it. There are reports of overexposure of workers to radiation and toxins in the plant.

There have been accidents at two of the heavy water plants which supply coolant material to the reactors. In 1977, a major explosion during a trial run forced a four-year shutdown of the Baroda plant. Another, at Talcher in Orissa, was rocked by an explosion and a major blaze in the summer of 1986 which sent hundreds of panicked residents fleeing. No fatalities were attributed to either incident.

Problems have been reported from Kalpakkam and also from Dhruva, the research reactor at Trombay, across the bay from Bombay. In two separate incidents, the dumping of nuclear waste from the fabrication complex in Hyderabad killed at least six persons, including two children, in 1980 and 1982. The victims handled highly combustible zirconium scrap. The dumping ground was, at the time, accessible to outsiders.

After Chernobyl, the country's nuclear plants were scrutinized by official agencies. At a news conference in November 1986, Dr Raja Ramanna, the chairman of the Atomic Energy Commission, and his senior associates declared that safety containment procedures in the country's nuclear reactors would not allow a Chernobyl-type incident to occur in India.[40] But to make sure there was no room for doubt the AEC chief said officials were developing disaster preparedness and emergency evacuation plans.[41] Ramanna was asked about Narora, where an atomic power plant is being set up, despite concern that the facility is located in an earthquake-prone zone and on the banks of the Ganges. His response was that they were building 'double containment' which would ensure that the facility was secure against seismic activity.[42]

Despite these assurances, there are many sceptics who are unconvinced about the nuclear programme. They do not believe it is safe enough.

Agriculture has been one of the showpieces of India's economic growth, despite the relatively low investments the Government has made in it. After years of being a major importer of foodgrains, India has finally become an exporter. In 1966-67, the years of the great Bihar famine, the country imported thirteen per cent of its foodgrains. In 1985 before drought conditions set in after the monsoons failed, India sent 150,000 tons of wheat to Ethiopia to help tide over a famine there.

Unfortunately, when considered in the light of the environmental problem, this growth in agricultural production has also led to a rise in the use of the country's pesticides. Between the 1950s and 1984, pesticide production soared from about 154 tons to 78,000 tons. (Yet, the country still loses nearly twenty per cent of its food production to pests). In 1984, the Government appointed a senior committee to evaluate the role of D.D.T. and BHC in the country's agricultural development. More recently, it expanded the scope of this committee to study whether pesticides which were banned or strictly restricted elsewhere in the world should be used here.

There are 119 pesticides registered under the Insecticide Act, with 6,500 different formulations.[43] Many of them are used indiscriminately by farmers who concoct cocktails or local mixes of different pesticides because they are improperly tutored in pesticide use. One result is that there are high residual levels of pesticides, both in the soil and in the harvested crop.

The CSE says that, in terms of volume, seventy per cent of all pesticides used in India are banned or restricted in the West.[44] The list of 'restricted' pesticides includes D.D.T., BHC, said to be twice as toxic as D.D.T., and methyl parathion, 20 times more dangerous than D.D.T. These are the main ones. Then come Herbicide 2,4-D (an active ingredient of Agent Orange, the defoliant used in Vietnam by US troops); DBCP(dibromochloropropane), which is banned in the United States for causing stomach cancer and infertility but is used on wheat in India, and phosvel which is also banned in the United States. As many as eighty pesticides banned in the West, including aldrin, dieldrin, heptachlor and endosulfan, are imported regularly in different quantities, says N.C. Joshi, director of the Central Plant Protection Training Institute at Hyderabad.[45]

In recent years, the Government has responded more quickly to media criticism of the use of such pesticides. Under an ongoing cooperation programme in agriculture, New Delhi has asked the United States to provide it with updates on the toxicity of numerous pesticides and their long-term effects.

However, there are few checks on how farmers and labourers actually apply the pesticides. Studies show that farmers spray more than the recommended use in the mistaken belief that they must kill all pests and that heavier use means fewer pests. 'Indiscriminate use of pesticides is seen throughout India', says Shannon Wilson, a US agricultural expert. Medical specialists say agrochemical workers who use and spray pesticides have reported vision damage, dislike of bright lights and night blindness. They say that studies have shown that users and victims of pesticide poisoning suffer from depression, insomnia and anxiety.

Those who have been affected the most are said to be workers in indigenous pesticide manufacturing industries. There are an estimated 774 pesticide factories and formulation units in India, with 712 in the small scale sector. (It is interesting to note that despite the proliferation of small scale units, MNCs continue to dominate production. Major MNC producers include Bayer, Union Carbide, Monsanto, Ciba-Geigy and Stauffer Chemical; these generate about one-third of the total manufacture). The law requires that all workers wear protective clothing and gear like masks, helmets, gumboots and gloves. More often than not, these stipulations are ignored.

The newsmagazine *India Today* reported on conditions at a formulation plant near Madras, describing it as indicative of the situation prevailing in the industry. The magazine interviewed a worker named Mohan, 'Half-naked, barefooted and coated with sulphuric dust.' The worker said, "I get severe cough and bronchitis every now and then and my health is poor for my age. I don't know how long I can last out. I can't afford to consult a doctor, and in our factory also there are no facilities for medical treatment'."[46]

According to C.R. Ramachandran, an occupational health specialist in Delhi, such cases are common. He says that pesticide users, including agricultural sprayers sent out by the Government, 'wear no masks, use no protective clothing, do not look to see if they are downwind or upwind'.[47] He says there will be little improvement in present conditions until knowledge of the dangers of the indiscriminate use of chemicals reaches workers in the field. He adds that this will not happen until there is higher literacy and better incomes among workers.[48]

In 1980, the Indian Institute of Management at Ahmedabad reported that fifty per cent of all pesticide users did not use protective clothing and about twenty per cent did not even wash their hands after the use of pesticides.[49] Of those who did, eighty per cent did not use soap.[50] Why so? Manufacturers say that skilled labour is difficult to find and one formulator even maintains that 'the effort at educating workers to use gloves, masks and boots is a waste. Moreover, they consider these as obstacles in their work.'

In the late 1960s, the Central Insecticides Board announced that all pesticides in use until then would need to be registered. The Insecticide Act was passed in 1971 and the 400 formulations in use then were asked to be submitted to the Board for analysis. Some manufacturers obliged but the majority did not furnish any data at all. Even if they had it is doubtful that it would have done much good. Some of the testing laboratories—there are a total of twenty-nine—are in poor condition. Even those that have some equipment are burdened with faulty machinery.[51]

Indians have the dubious distinction of having perhaps the world's highest levels of accumulated DDT in their body tissues: between 12.8 ppm and 31 ppm.[52] Yet compare this with the per hectare consumption of pesticides in the country: 180 grammes. This figure per hectare is far lower than the average of 1,490 grammes in the United States and 10,790 grammes in Japan. What explains the heavy accumulation of DDT is probably the indiscriminate use of toxic insecticides before harvesting.

Other strong crop medicines used just before sending the vegetables to the markets were identified as BHC, aldrin, folidol dust and endosulfan. While the residual effects of endosulfan lasts up to fifteen to twenty days, BHC, aldrin and folidol dust are highly persistent. A report by the Himachal Pradesh agricultural university said these insecticides should be used 'at the time of transplantation or before sowing', and that farmers should not spray the drugs at harvest time.

Two scientists from the Punjab Agricultural University at Ludhiana did a study of pesticide residues in cattle milk and found that all samples from the countryside showed DDT residue above the tolerance level. A WHO study reported that fifty per cent of the cereals, milk, meat, eggs and pulses samples it had studied from across India contained pesticide residues. About thirty per cent of the samples contained levels above tolerance limits.[53]

The Indian Agricultural Institute in New Delhi found that vegetables coming into the capital contained insecticide residues which were twenty times the permissible limit.[54]

The failure of farmers to follow the simplest of instructions is seen not just in India but all across the developing world. The Pesticide Poisoning Report of the International Organization of Consumer Unions lists case after case in country after country where there is blatant, indiscriminate use.

The reasons for this, as with India, are illiteracy, lack of comprehension of safety requirements and inadequate instruction in the use of pesticides.

Analysis and common sense too point to one other basic fact: that insects and pests become resistant to pest control measures. Hence, farmers must use more and more pesticides to deal with newer and stronger enemies in the field.

In 1954, there were only twenty-five species of arthropods which were resistant to insecticides. By 1980, that figure had risen to 432. Consequently, farmers in Gujarat spray their fields twenty or thirty times more often than before. Maharashtra's Vidharba region, one of the most arid in the country, has reported a three-fold rise in pesticide use. From Andhra Pradesh, comes the news that at least 15 species of pests are now resistant to all commonly used agrochemicals. Andhra Pradesh is the heaviest consumer of pesticides in India, accounting for about one-fifth of total use.[55]

'The industry gives work, even if it is dangerous, and that is what seems to count for most people', says Arun Kumar, an economic analyst.[56] 'People are more concerned about food now, jobs now rather than later, instead of worrying about the possible health effects of what they are using ten, fifteen, twenty years from now'.

At the risk of being accused of being oversimplistic, it must be stressed that everything in India—Bhopal, Antop Hill, indiscriminate pesticide use, slums around industry, lack of sanitation, millions below the poverty line—boils down to a matter of people. India has too many of them. And there will be far more by the end of the century than the Government can adequately provide for. According to some projections, at the current rate of growth, there will be over one billion Indians in the subcontinent by the turn of the century. Of this figure, more than half, or about 600 million will still live in abysmal poverty, their families earning less than Rs. 6,400 per year, the current cut-off figure for the poverty line. The first priority of the Government, therefore, is a strategy to regulate population growth. This does not mean that other areas such as planned agricultural and industrial growth, the necessity to check pollution and enhance industrial safety should be neglected. But the tragedy at Bhopal showed that when there are too many people, many more die. This was the most obvious aspect. What wasn't so apparent was that everything was linked to the millions burdening every aspect of India. The pesticide plant existed to enhance the quantities of crop, which in turn went to feed the people. The people lived next to the plant because there were too many of them to make a living off traditional methods of agriculture in the countryside. More tragically, they lived next to the plant because they had to. There was simply no other place to go.

From the general to the specific. While there might be many causes for what occurred at Bhopal, one thing was clear: industrial safety could no longer be taken lightly. And if there was any welcome outcome of the tragedy at all, it was that there appears to be a new awareness in industry and government of the need for industrial safety.

One industrialist reported at a seminar in Madras of how his factory's consumption of water had dropped from 40,000 litres per hour to 4,000 litres per hour after the installation of an effluent plant. Such expense, another industry leader said at the same conference, should be viewed 'as an essential price'.[57]

'It is better to act now, at a time of our choosing, rather than have the demands of an agitated society thrust on us', says Suresh Krishna, a vice president of the Confederation of Engineering Industry. 'Government legislation and statutory controls should be seen as deterrents for those who chose to ignore their responsibility'.[58]

# The Lessons Of A Tragedy

There are no categorical answers to the questions and issues raised by the events of December 2-3, 1984. The risks as well as the fruits of industrial and economic progress cannot be denied and will remain with us. But an important step forward would be the proper management of economic growth so that nations, corporations and people can balance the benefits of technology with its dangers.

Hazardous industries cannot be shut down. It would be naive even to consider such a step. But the Bhopals, the Sevesos, the Minamatas and the Basels have led to a situation where large corporations are being forced, by the weight of their collective sins, to adopt tighter measures of safety. This is one positive outcome of the various tragedies. The other is that growing public awareness in Europe, the United States, India and other Asian nations, such as Indonesia and the Philippines—where governments have yielded to the strength of public indignation against the disinformation and double standards of transnationals—has contributed to new and more comprehensive legislation, and various other initiatives to ensure the safe operation of hazardous industry. Yet only a beginning has been made and there can be no question that it will be a long time, if ever, before an all-encompassing safety code can be developed and, more important, implemented by those who are expected to abide by it—the multinationals and other sectors of industry. It would be hypocritical if such a code of conduct did not cover local entrepreneurs and large, medium and small-scale indigenous industries who are perhaps more involved in the development processes in their respective countries. There is no doubt that industry would resent being regulated but it is in its own interest to regulate itself and also be more open.

Governments too must become more open about their shortcomings, instead of trying to pretend that they can do no wrong. The governments of developing countries in particular need to frame industrial policies that are fair and equitable while also insisting on stringent safety measures and guarantees. It is in making industry safe yet profitable that true economic growth and the well-being of a nation's people lies. Governments will need to give special incentives to industry which will make compliance with safety rules an attractive proposition. At the same time, they must have tough enough laws, and the political will to enforce them.

Environmental groups have a key role to play, as watchdogs, especially in societies where the spread of education and the awareness of industrial hazards is low. That way both industry and government will, perforce, be kept on their toes.

Activist groups have a second role: maintaining and sustaining a minimum level of public interest in the twin issues of environmental protection and industrial safety.

Universities and schools have a part to play here. They need to encourage awareness about the environment among their students. Enlightened multinationals and local industrial houses could play a practical part in the development of a national consciousness by funding education programmes that seek to impart such knowledge.

That said, in order to achieve in practice what is so often limited to theory, there can be no substitute for the battle for mankind's safety beginning at the grassroots level.

Karnataka state's S.G. Hiremath is an outstanding example of what the ordinary man can achieve.[1] He has successfully initiated public agitations and legal suits against both private industry and government and literally forced them to take notice of what needs to be done.

Since the 1970s, Hiremath has worked in Dharwar district organizing villagers against pollution from private and public sector firms. He was first prompted to act when he saw untreated effluents pouring out of a Birla factory (the Birlas are among the biggest of India's industrial houses). The discharge was causing massive fish kills in the Tungabhudra river, polluting the river water and impairing the health of villagers who depended on the river for food and drinking water.

When appeals to plant managers and government officials were ignored, Hiremath and local activists launched rallies and demonstrations. The protests drew wide publicity in the state and national media. Bowing to pressure, the company installed pollution control devices and cleaned up its operation. The state pollution control board inspected the factory and monitored its effluents. Finally, in an unprecedented gesture, Hiremath was appointed to the state pollution control board as a representative of the public interest.

'This is the only way to force the hand of industry and government,' says Hiremath, 'public pressure and agitation backed by media coverage; they're scared of publicity, especially bad publicity.'

He emphasizes the fact that the leadership and initiative must be locally-based, not imported from another area. 'The local people must feel involved and feel they are leading a movement which is protecting their interests,' he says.

Hiremath feels a major drawback of the activist effort at Bhopal was that its organization was in the hands of non-Bhopalites. The people of the city, he believes, never really felt they had a stake in the work of the activists.

If Hiremath's example is something that can be held up to other activist groups, then there are precedents too where multinationals have been made to toe the line in the developing world. The Indian Government's prolonged, painful effort to bring Union Carbide to justice is not a particularly inspiring example. What is required is swift, clear, affirmative action, as was seen in Bangladesh in 1982.

That year, Bangladesh announced a new drug policy aimed at eliminating harmful and useless medicines and increasing the domestic production of essential drugs.[2] It set up a public distribution system for such essential drugs and ordered the use of generic rather than brand names.

Multinational corporations were told they would not be allowed to produce antacids and vitamins, two of their staple items in the country. The Government said this would leave them free to concentrate their efforts and resources on those items not so easily produced by smaller national companies. Cough mixtures, throat lozenges, gripe water and alkalis were banned. The World Health Organization says that as many as 4,000 drugs were sold before the new policy. Nearly 1,700 of these were 'either useless or harmful.' The new essential drugs list contained only 150 items (of these, twelve are used by village-level health workers to treat common ailments).

The policy, predictably, created an uproar, especially among transnationals with production facilities and agencies in Bangladesh. Still they had no option but to comply with the new rules.

The prices of essential drugs dropped. The policy is still being enforced. And multinationals are still doing business in Bangladesh.

For years there was has been talk of a code of conduct for multinationals. However, nothing acceptable to all sides has yet been evolved. One reason for this is that governments are unable to shed their fascination for the carrots that multinationals dangle before them. Should they go for the lucre or should they opt for planned, sensible, safe development? Most governments try to do both but find it difficult to strike a balance.

Five basic points should form part of any eventual code. These are:

1. Safety levels and equipment in a subsidiary must match those in the parent firm.
2. MNCs must share with the relevant government agencies all data relating to the toxicity of materials used in the production process, especially any recorded effects on humans, animals or plants and antidotes for accidental emergency spills/other emergency situations. These must be made public through the news media at times of crisis.
3. MNCs must agree to settle any reasonable compensation claim for damage and injuries caused by an incident originating in the multinational or its subsidiary plant.
4. All MNCs and their subsidiaries must develop emergency

evacuation and disaster preparedness plans with the assistance of local regulatory agencies, state and central administrations, local neighbourhood leaders and public interest groups.

5. All the points mentioned earlier must form part of the submissions that MNCs make before local administrations for permission to start new units. Only after these various aspects have been assessed and found adequate should MNCs be allowed to initiate new ventures. These conditions should be made binding on indigenous industry as well.

Apart from underlining the need for big industry to adopt the most stringent safety standards and for governments to ensure they do so, Bhopal also stressed the need for the following:

* Governments and industry must set higher standards for pesticide testing and use. Field workers and farm labourers must be carefully instructed in the proper use of pesticides and allowed to handle them only after understanding the risks and the need for protective equipment. Some officials suggest the setting up of National Poison Centres where medical specialists and toxicologists supported by accurate, sophisticated equipment, can draw on a data bank to deal with overexposure to chemicals.

* MNCs and local industry must play a responsible role in developing such data banks and poison centres. They can fund, staff and furnish them. But the work should be assessed by independent audits so that research work is not diverted into secret, unauthorized fields.

* Local administrations must enforce zoning laws and prohibit population centres around toxic industrial units. But they must be given the flexibility to permit communities around non-toxic plants, again within specified limits. Slum dwellers around industrial areas must be notified of the hazards and involved in emergency drills and safety procedures.

* Governments must increase incentives for industries to grow in rural, backward areas so that the exodus to the cities is slowed.

* Politicians who encourage slums around hazardous plants

must be declared offenders and punished by the law. Political parties must develop policies on the environment.

* Industry and governments must encourage and fund the search for alternative technologies to chemical pesticides. Work is continuing on the sterilization of pests, as an alternative to bombarding them with chemical weapons. Other natural forms of pest-control, such as using natural predators, could help restore some of the delicate checks and balances that nature has woven and man, in his arrant stupidity, has demolished. Much more needs to be done in the critical field of biotechnology. This is a developing science which uses germ warfare on insects, instead of chemical weaponry. It involves the use of fungi, bacteria and viruses which infect and destroy insects. Four corporations, including Sandoz, Monsanto and Abbott, sell microbes as pesticides.[3]

These are still too expensive to be practical but they could eventually be in the vanguard of means to rid the world of unnecessary poisons.

* Organic farming, though expensive and traditional, using cow dung and other manure, needs to be supported. This should reduce the persistence of toxic pesticides in the soil and in food. In countries like India, where there are plenty of cattle and consequently a lot of dung, imaginative plans need to be formulated for its use.

* Different arms of the Government and regulatory agencies, often compete with each other instead of coordinating their work. There is much to be said for reducing the proliferation of government agencies and winding up those which duplicate the work of others. Those that do exist should be given the means to do their jobs more effectively.

But these suggestions will come to naught if governments and political parties lack the will to act. The present attitude of governments, especially in the developing world, is far from satisfactory. Despite Bhopal, after the initial brouhaha, no political party in India has yet made care for the environment or safety in hazardous industrial processes a top priority.

To take one example to illustrate this point.

In April 1987, Anil Agarwal, the well-known environmentalist, was invited to speak to a joint assembly of all Members of Parliament on the problems caused by flooding and excessive erosion.

First, the man who initiated the meeting didn't even turn up for it. That was Prime Minister Rajiv Gandhi. And only forty members of Parliament, barely eight per cent of the Lok Sabha or Lower House, put in an appearance. When they noticed the premier's absence, says Agarwal, the legislators also began to drift away.

When he completed his talk, there were only nine members of Parliament in the room.

That absymal level of political interest and support means that Indians—and other Third World nationalities—should not expect their governments to take a lead in issues of community safety, particularly when they involve the environment. The political will and effort appears to stop at the framing of laws. This needs to change and rapidly. And the only way it will is by the people putting pressure on their Government, whether local of national, as people like Hiremath, the Karnataka activist, do.

So finally it all comes down to people and their attitudes towards technology, the way it remakes their lives and the lives of those around them. People should want to know all the time what it is that cushions them and, at times, can kill them. It deserves to be repeated that attitudes need to change as much as the laws of the land.

Nations and people need to develop the 'hard hat' culture, widely practised in the West, which emphasizes safety. Without this attitude, no amount of new technology and new legislation can save people from the horrors of events like Bhopal. Public concepts of safety and cultural responsibility are way behind the new technologies spawned by modern science and they need to catch up. 'The advance of technology', said West German President, Richard von Weizsäcker, 'time and again confronts us with the question of whether we have a moral command on our technical ability'.[4]

The examples of Bhopal, the polluted Rhine, of Minamata and

Seveso, Shriram and Antop Hill must make us reconsider this question.

Moral responsibility begins where industrial managers and trade unions draw the legal line, where governments seek to hide their failures and where public activist groups fall short.

Eternal vigilance is the price nations pay for their liberty. They cannot be less vigilant about safety in the new development processes.

In 1928, author and essayist Henry Beston wrote: 'When the Pleiades and the wind in the grass are no longer a part of the human spirit, a part of very flesh and bone, man becomes, as it were, a kind of cosmic outlaw, having neither the completeness and integrity of the animal nor the birthright of true humanity.'

Others have put it less well. Progress cannot afford more Bhopals.

# Sources And Notes

*Introduction*


1. Harold J. Corbett, Senior Vice-President of the Monsanto Company, the *New York Times,* March 27 1985
2. *Down to Business: Multinational Corporations, the Environment and Development,* Charles S. Pearson, World Resources Institute, Washington, January 1985
3. The Fortune 500 list of top American companies
4. Ibid
5. Ibid
6. The World Bank Atlas 1985
7. Ibid
8. Ibid
9. Ibid
10. Fortune 500
11. World Bank Atlas
12. Ibid
13. Ibid
14. *Multinationals Under Fire: Lessons in the Management of Conflict,* Thomas N. Gladwin and Ingo Walter; Published by John Wiley and Sons
15. Ibid
16. Various news accounts
17. US Senate, 94 Congress: Staff Report of the Senate Committee to Study Government Operations with Respect to Intelligence Activities, Washington, 1976; Appendix AI, Covert Action in Chile 1963-1976
18. The *New York Times,* 23 December 1976
19. *The United States and Chile—Imperialism and The Overthrow of the Allende Government,* James Petras and Morley,

Monthly Review Press, London, 1975, Quoted in Political Dimensions of Multinationals in India, Indian Institute of Public Administration, New Delhi, August 1983

20 Ibid

21 *Silent Spring,* Rachel Carson, Pelican Books

22 *Environment,* April 1985, Defining Hazardous Wastes, Michael Dowling

23-25 Ibid

26 *World Health,* August-September 1984

27 *Down to Business*

28 Ibid

29 *World Health,* August-September 1984

30 *Down to Business*

31 Ibid

32 *The Lessons of Bhopal,* Martin Abraham, International Organization of Consumers Union, Penang, 1986

33 Ibid

34 The *New York Times,* 18 November 1985

35 The *New York Times,* 25 November 1985

36 The *New York Times,* 27 March 1985

37 Ibid

38 Ibid

39 *Multinationals Under Fire,* The Seveso Incident

40-43 Ibid

44 *Circle of Poison,* David Weir and Mark Shapiro, The Institute for Food and Development, San Francisco

45 Ibid

46 Ibid

47 *Environment,* January-February 1985: Local and Multinational Corporations, Reappraising Environment Management, by Michael G. Royston

48 Speech to the International Labour Organization, Geneva, 17 June 1985

49 *The Multinationals,* Christopher Tugendhat, Penguin Books, Metheun London and Random House.

50 Ibid

51 Ibid

52 *Technology and Third World Multinationals:* Louis T. Wells, Working Paper No. 19, 1982, International Labour Office, Geneva

53 Ibid

54 *Down to Business*

55 John Noble Wilford, *International Herald Tribune*, 6 May 1986

SECTION I

*Out Of The Evil Night*

1 The interviews with Suman Dey and others quoted in this chapter in connection with the disaster were conducted by the author in July, August and September 1986. Some were interviewed in December 1984. All the interviews related to the incident were conducted by the author, unless specified otherwise

2 Interview with Vijay Gokhale, managing director, Union Carbide India Limited, August 1986

3 The *New York Times,* 19 May 1985: The personal ordeal of Warren Anderson by Stuart Diamond. The details about Mr. Anderson on this page are drawn from that article

4 *Fortune,* 28 April 1986, News and Trends

5 Jackson Browning's remarks were made at the 20 March 1985 news conference at Danbury

6 *Environment,* September 1985: *Avoiding Future Bhopals,* B. Bowender, Jeanne X. Kasperson and Roger Kasperson

7 The *New York Times,* 13 August 1985

8 Ibid

9 *No Place to Run,* Highlander Center and Society for Participatory Research in Asia, May 1985

10 The Trade Union Report on Bhopal, The International Confederation of Free Trade Unions, Geneva, July 1985

11 Union Carbide, Volunteer Workers Exposure Monitoring Study (Panama). Report of Union Carbide Corporation Submitted to the World Health Organization (unpublished)

12 *No Place to Run*

13 The *New York Times,* 10 July 1985

14 *No Place to Run*

15 Ibid

16-17 *The Asian Wall Street Journal,* 3 June 1986

18 This and subsequent figures for UCIL's financial condition are based on the company's annual reports for 1983 and 1984

19 Ibid

20 Statement by Union Carbide Corporation before the Bhopal District and Sessions Court, November 1986

21 Company documents
22 Testimony by Warren Wommer in pre-trial discovery hearings in the United States
23 Ibid
24 Affidavit by Edward Munoz
25 Ibid
26 Remarks by James H. Rehfield to lawyers for India and the corporation in response to questions at a pre-trial discovery hearing in New York
27 Trade union report
28 Ibid
29 Union Carbide safety survey of UCIL, Bhopal, May 1982
30-32 Ibid
33 Affidavit by Union Carbide Corporation, Bhopal, October 1986
34 Government of India affidavit at Bhopal, November 1986
35 Report on scientific studies on the factors related to Bhopal toxic gas leakage, December 1985
36 Trade union report

### *The Fight to Save Lives*

1 *New Scientist,* 20-21 December 1984: also *Chemical and Engineering News,* 11 February 1985
2 *Chemical and Engineering News,* 11 February 1985
3 *Health Effects of Bhopal Gas Tragedy,* April 1986, Indian Council of Medical Research
4 Ibid
5 *The Excess of Gynaecological Diseases in Women Exposed to Methyl Isocyanate Gas in Bhopal,* 1985 Rani Bang
6-8 Ibid
9 Interview with author, July 1985, Bombay
10-12 Ibid
13 Medical survey on Bhopal gas victims conducted by a team of doctors and technicians from Bombay Municipal Corporation and specialists from the King Edward Hospital, Bombay, May 1985
14 Interview with author, July 1985
15 Ibid
16-17 The section on Dr. Kamat's work is based on interviews with the author in July 1985 and July 1986

18-19 *Journal of Postgraduate Medicine,* 1985. *Early Observations on Pulmonary Changes and Clinical Morbidity due to the Isocyanate gas leak at Bhopal,* by S.R. Kamat
20 Ibid
21 Interview with author, July 1986. Subsequent quotes from Ishwar Dass in this chapter are from this conversation.
22 Confidential note to Mr Vora, January 1986. The official who wrote the note has requested anonymity. This letter figures in my notes
23 Ibid
24 Ibid

## *The Legal Battle*

1 The *Washington Post,* 14 December 1984 (Reprint in the *International Herald Tribune*)
2 *The New York Times,* 18 December 1984
3 Interview with author, Bhopal, December 1984
4 Ibid
5 *India Abroad,* New York, 25 April 1986
6-10 Ibid
11 Interview with author, Bhopal, July 1986. Khanna's remarks in this chapter are from this interview unless otherwise indicated
12 The *New York Times,* 12 March 1985
13-14 Ibid
15 Affidavit of Union of India, U.S. District Court, South district of New York, 8 April 1985
16-20 Ibid
21 The *New York Times,* 17 April 1985
22 The *New York Times,* 19 April 1985
23 Interview with reporters including author, New Delhi. 19 April 1985
24 The *New York Times,* 25 April 1985
25 Ibid
26 The *New York Times,* 20 June 1985
27-29 Ibid
30 Memorandum of Law in support of Union Carbide Corporation's motion to dismiss these motions on the grounds of *forum non conveniens,* New York, 31 July 1985
31-42 Ibid

| | |
|---|---|
| 43 | Memorandum of private plaintiffs in opposition to Union Carbide Corporation's motion to dismiss the case on grounds of *forum non conveniens* |
| 44 | Ibid |
| 45 | Ibid |
| 46 | Ibid |
| 47 | Ibid |
| 48-50 | Rulings in Manu International, DS.A. vs Avon Products Inc. |
| 51-55 | Union Carbide motion for dismissing cases |
| 56-60 | Affidavit of Nani A. Palkhivala in support of Union Carbide motion. 18 December 1985 |
| 61-68 | Affidavit of Marc S. Galanter, 5 December 1985 |
| 69-72 | Memorandum of Law in opposition to Union Carbide Corporation's motion, Plaintiffs Executive Committee, 6 December 1985 |
| 73-77 | Ibid |
| 78 | The *New York Times,* 10 December 1985 |
| 79-85 | This is based on various news accounts and court documents |
| 86-91 | Notice from Michael V. Ciresi to the Board of Directors of Union Carbide Corporation, 23 December 1985 |
| 92 | Hagler quoted by Shahnaz Anklesaria Aiyar, *Indian Express,* 31 March 1986 |
| 93 | Ibid |
| 94-97 | Brief *amicus curiae* of Citizens Commission on Bhopal, The National Council of Churches, The United Church of Christ, Commission on Racial Justice, et al on *forum non conveniens* |
| 98-100 | Interview with Indian Government official, June 1986 |
| 101-102 | Interview with Dr. N.P. Mishra, Bhopal, July 1986 |
| 103 | The *New York Times,* 23 March 1986 |
| 104 | Interview with Indian Government official present at meeting |
| 105-106 | Interview with Indian official who quoted the Ciresi letter, June 1986 |
| 107 | Justice A.N. Mulla report on the condition of prisons in India |
| 108-110 | *Indian Express,* 4 August 1986 |
| 111-112 | Interviews with relief officials in Bhopal, July 1986 |
| 113 | Interviews with the author, July 1986 |
| 114-123 | Affidavit on behalf of Union Carbide Corporation, Bhopal, November 1986 |

*Who Was Responsible?*

1 Quoted in Union of India's reply to Union Carbide statement alleging sabotage, January 1987. Also quoted earlier in 6 December 1985 motion before Judge Keenan to dismiss Union Carbide motion of *forum non conveniens*.

2 Quoted in 1985 motion to dismiss Union Carbide plea. Telex was dated 13 November 1984, three weeks before the Bhopal disaster

3-4 Internal correspondence, 11 September 1984. Engineering and Technology Services/Central Engineering Department, Safety Health Group, Union Carbide Corporation

5-11 Ibid

12 Bhopal Methyl Isocyanate Incident Investigation Team Report, Union Carbide Corporation, March 1985

13 Interview with Labour Ministry officials, July 1986

14 Law passed by State Government in April 1984, conferring tenancy rights on all illegal *jhuggi jhopri* (shanty town) dwellers in the state

SECTION II

*A Poisoned World*

1-2 *Gluttons for Punishment,* James Erlichman, Penguin Books, 1986

3-4 Ibid page 29

5 This and subsequent reference to chemical spills and their aftermath in Europe are based on reports published in the *International Herald Tribune,* November 1986

6 References to the Kepone incident in Hopewell, Va. are based on an account published in The *Richmond Times-Dispatch,* 9 June 1985

7 The *New York Times* 27 October 1986. This and subsequent references to Chernobyl are based on that report.

8 This and subsequent references to the Seveso episode are based on an account published in *Multinationals Under Fire: Lessons in the Management of Conflict,* Thomas N. Gladwin and Ingo Walter, John Wiley and Sons

9 Remarks at Union Carbide news conference at Danbury, 29 March 1985

10 This and subsequent references to the Seveso directive on this page are based on the Council's directive of 24 June 1982 on major accident hazards of certain industrial activities.
11-14 April 1983 bulletin of the European Community
15 Official journal of the European Community, 16 May 1983
16-23 *The Bhopal Syndrome,* David Weir, International Organization of Consumers Unions, Penang, 1986
24 *The Pesticide Poisoning Report:* A Survey of Some Asian Countries, International Organization of Consumers Unions, Penang, 1985
25-28 *Multinational Monitor,* December 1985, *The Dirty Dozen: Pesticides we can do without*
29 *Gluttons for Punishment,* James Erlichman, Penguin Books
30-31 *Fear in Unknown Quantities,* Drew Douglas, March 1985
32-33 *Silent Spring,* Rachel Carson, Metheun London and Penguin Books
34-36 *Fear in Unknown Quantities* Douglas
37-39 *An Immodest Proposal,* Sierra, Carl Pope September/October 1985
40-42 *Environment,* January/February 1985: *Local and Multinational Corporations: Reappraising Environment Management,* Michael G. Royston
43-46 Pope, Sierra, September/October 1985, op cit
47 This and subsequent references to the two studies of Multinational Corporations are based on Royston's report in *Environment,* op cit
48 This description of the Minamata disaster is based on *Minamata Disease, Chronology and Medical Report,* by Masazumi Harada, M.D., of the Department of Neuropsychiatry Institute of Constitutional Medicine, Kumamoto University
49 Raj Kumar Keswani, *Illustrated Weekly of India,* 27 April 1986
50-51 Dr. Harada, op cit
52-53 Keswani, *Illustrated Weekly of India,* op cit

### *India's Neglected Environment*

1 This and subsequent references to M.C. Mehta in this section are based on interviews by the author, July/August 1986 and January 1987

| | |
|---|---|
| 2 | The account of the accident at Shriram Food and Fertilizers Limited and its aftermath is based on news reports in various Delhi newspapers |
| 3 | Company report, 1984 |
| 4 | The *Indian Express,* 14 December 1985 |
| 5 | Interviews with company officials, July 1986 |
| 6 | Interviews with the author, June 1986. Subsequent references to Mr. Chowdhury are based on these interviews |
| 7 | Supreme Court ruling, 17 February 1986 |
| 8 | Supreme Court Judgement, 20 December 1986 |
| 9 | Interview with author, January 1987 |
| 10 | Ibid |
| 11 | Ibid, May 1986 |
| 12-14 | *Business India,* 14-21 July 1986 |
| 15 | Report of Inter-Ministerial Group, *Safety in Chemical and Petrochemical Industries,* September 1986. Subsequent references to chemical industry on pages 267 and 268 are based on this report |
| 16-17 | Interviews with officials of the Directorate-General, Factory Advice Services and Labour Institute, Bombay, August 1986 |
| 18 | Environment Safety Committee Report 1985-86, Maharashtra Pollution Control Board |
| 19 | *Bombay* 7-21 April 1985 |
| 20 | Centre for Science and Environment, *The State of India's Environment,* 1984-85 (section on Health). This and other details about health hazards faced by Indian Industrial Workers are based on the CSE Report |
| 21-22 | *Newsday,* 16 December 1981 |
| 23-26 | Environment Safety Committee Report, 1985-86 |
| 27 | Interview with *D-G FASLI* Officials, August 1986 |
| 28 | Standard reference notes on activities of the *D-G FASLI* for 1985. This and subsequent information relating to and interviews with officials at D-G *FASLI* |
| 29 | Report of the Inter-Ministerial Group, September 1986 |
| 30-31 | *Environmental Guidelines on Siting of Industry,* Report of the Working Group, August 1985, Ministry of Environment and Forests |
| 32-34 | Interview with author, August 1986 |
| 35 | Ibid, July 1986 |
| 36 | *Business India,* September 1986 |
| 37-38 | Interview with author, October 1986 |
| 39 | The Material on the Indian nuclear industry is based on *The* |

*State of India's Environment, CSE* 1984-85

40-42 Press conference by Dr. Raja Ramanna, New Delhi, December 1986

43 N.C. Joshi, Distribution and use of Agro-Pesticides in India, Central Plant Training Institute, Hyderabad (undated)

44 *The State of India's Environment, CSE,* 1984-85

45 Joshi, op cit

46 *India Today,* 15 March 1986

47-48 Interview with author, September 1986

49-50 *India Today,* 15 March 1986

51 Interview with official at *DG-FASLI*

52 R.L. Kalra and R.P. Chawla, Punjab Agricultural University, Ludhiana, Monitoring of Pesticide Residues in the Indian Environment (1980)

53-55 *State of India's Environment, CSE,* 1984-85

56 Interview with author, September 1986

57-58 *The Hindu,* 30 October 1986

*The Lessons of a Tragedy*

1 Interview with author. The account of Mr. Hiremath's work is based on this interview, September 1986, New Delhi

2 *World Health* July 1984. The account of Bangladesh's campaign for the use of essential drugs is based on a report in *World Health Magazine.*

3 *Gluttons for Pubishment,* James Erlichman, Penguin Books

4 8 April 1986 Speech at the Hanover Industrial Fair

## MORE ABOUT PENGUINS

For further information about books available from Penguins in India write to Penguin Books (India) Ltd, Room 2-4, 1st Floor, PTI Building, Parliament Street, New Delhi-110 001.

*In the UK* : For a complete list of books available from Penguins in the United Kingdom write to Dept. EP, Penguin Books Ltd, Harmondsworth, Middlesex UB 7 ODA.

*In the U.S.A.* : For a complete list of books available from Penguins in the United States write to Dept. DG, Penguin Books, 299 Murray Hill Parkway, East Rutherford, New Jersey 07073.

*In Canada* : For a complete list of books available from Penguins in Canada write to Penguin Books Canada Ltd, 2801 John Street, Markham, Ontario L3R IB4.

*In Australia* : For a complete list of books available from Penguins in Australia write to the Marketing Department, Penguin Books Australia Ltd, P.O. Box 257, Ringwood, Victoria 3134.

*In New Zealand* : For a complete list of books available from Penguins in New Zealand write to the Marketing Department, Penguin Books (N.Z.) Ltd, Private Bag, Takapuna, Auckland 9.

# COLLECTED POEMS (1957-1987)

*Dom Moraes*

Myth, religion, grief, friendship, war, love and alienation are some of the themes Dom Moraes, arguably India's greatest living English language poet, deals with in his poetry. Though his poems can be enjoyed for their lyrical beauty alone, on other levels they draw the reader into extraordinarily powerful atmospheres of various aspects of life—romance, passion, fear, old age, death and renewal.

In addition to the verse published in his three previous books of poetry—*A Beginning, Poems,* and *John Nobody—Collected Poems* contains over fifty unpublished poems.

'Dom Moraes' poems exhibit...a mellifluous smoothness, a sinuous fluency, a delicacy of mood and perception, a liking for regular metres and a gift for the vivid phrase.'

*The Sunday Times*